Alessandro Carbotti, Serena Dipierro, and Enrico Valdinoci

Local Density of Solutions to Fractional Equations

De Gruyter Studies in Mathematics

Volume 74

Alessandro Carbotti, Serena Dipierro, and
Enrico Valdinoci

Local Density of Solutions to Fractional Equations

DE GRUYTER

Mathematics Subject Classification 2010
Primary: 26A33, 34A08, 35R11; Secondary: 60G22

Authors
Alessandro Carbotti
Dipartimento di Matematica e Fisica
Università del Salento
Via Per Arnesano
73100 Lecce
Italy
alessandro.carbotti@unisalento.it

Prof. Dr. Enrico Valdinoci
Department of Mathematics and Statistics
University of Western Australia
35 Stirling Highway
Crawley, WA 6009
Australia
enrico.valdinoci@uwa.edu.au

Prof. Dr. Serena Dipierro
Department of Mathematics and Statistics
University of Western Australia
35 Stirling Highway
Crawley, WA 6009
Australia
serena.dipierro@uwa.edu.au

ISBN 978-3-11-066069-2
e-ISBN (PDF) 978-3-11-066435-5
e-ISBN (EPUB) 978-3-11-066131-6
ISSN 0179-0986

Library of Congress Control Number: 2019946148

Bibliographic information published by the Deutsche Nationalbibliothek
The Deutsche Nationalbibliothek lists this publication in the Deutsche Nationalbibliografie; detailed bibliographic data are available on the Internet at http://dnb.dnb.de.

Typesetting: VTeX UAB, Lithuania
Printing and binding: CPI books GmbH, Leck

www.degruyter.com

Preface

The study of nonlocal operators of fractional type possesses a long tradition, motivated both by mathematical curiosity and by real-world applications. Though this line of research presents some similarities and analogies with the study of operators of integer order, it also presents a number of remarkable differences, one of the greatest being the recently discovered phenomenon that *all functions are (locally) fractionally harmonic (up to a small error)*. This feature is quite surprising, since it is in sharp contrast with the case of classical harmonic functions, and it reveals a genuinely nonlocal peculiarity.

More precisely, it has been proved in [25] that given any C^k-function f in a bounded domain Ω and given any $\epsilon > 0$, there exists a function f_ϵ which is fractionally harmonic in Ω such that the C^k-distance in Ω between f and f_ϵ is less than ϵ.

Interestingly, this kind of results can be also applied at any scale, as shown in Figures 1, 2, and 3. Roughly speaking, given *any* function, without any special geometric prescription, in a given bounded domain (as in Figure 1), one can "complete" the function outside the domain in such a way that the resulting object is fractionally harmonic. That is, one can endow the function given in the bounded domain with a number of suitable oscillations outside the domain in order to make an integro-differential operator of fractional type vanish. This idea is depicted in Figure 2. As a matter of fact, Figure 2 must be considered just a "qualitative" picture of this method, and should not be regarded "realistic." However, even if Figure 2 does not provide a correct fractional harmonic extension of the given function outside the given domain, the result can be repeated at a larger scale, as in Figure 3, adding further remote oscillations in order to obtain a fractional harmonic function.

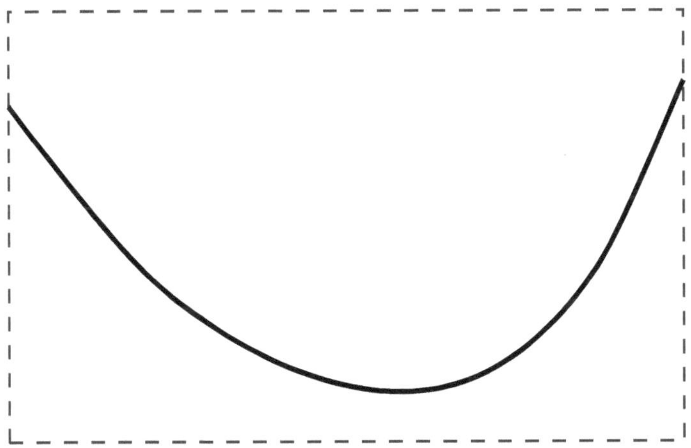

Figure 1: All functions are fractional harmonic, at different scales (scale of the original function).

https://doi.org/10.1515/9783110664355-201

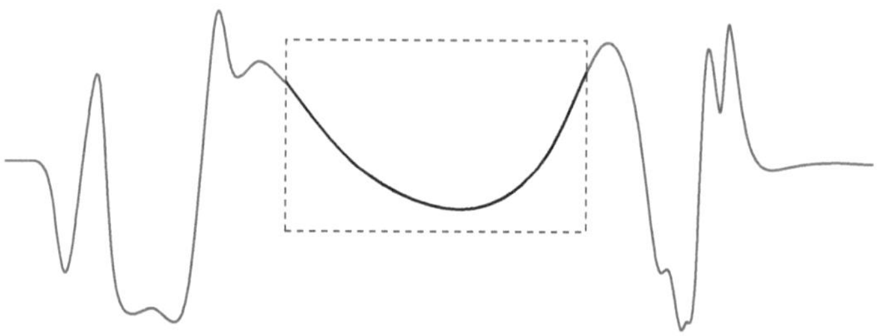

Figure 2: All functions are fractional harmonic, at different scales ("first" scale of exterior oscillations).

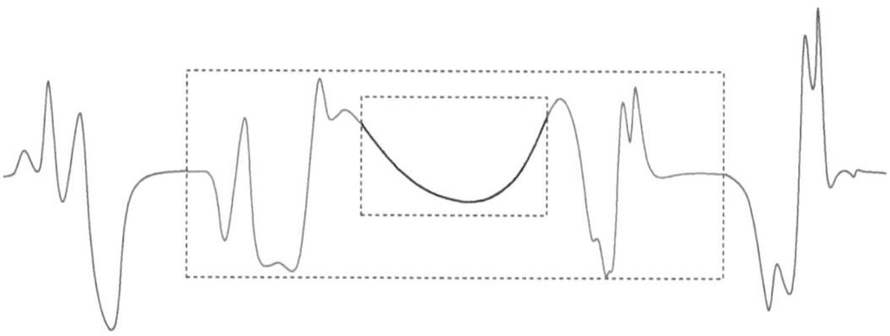

Figure 3: All functions are fractional harmonic, at different scales ("second" scale of exterior oscillations).

In this sense, this type of results really says that whatever graph we draw on a sheet of paper, it is fractionally harmonic (more rigorously, it can be shadowed with an arbitrary precision by another graph, which can be appropriately continued outside the sheet of paper in a way which makes it fractionally harmonic).

This book contains a *new result* in this line of investigation, stating that *every function lies in the kernel of every linear equation involving some fractional operator, up to a small error*. That is, *any given function can be smoothly approximated by functions lying in the kernel of a linear operator involving at least one fractional component*. The setting in which this result holds is very general, since it takes into account anomalous diffusion, with possible fractional components in both space and time. The operators taken into account comprise the case of the sum of classical and fractional Laplacians, possibly of different orders, in the space variables, and classical or fractional derivatives in the time variables. Namely, the equation can be of any order, it does not need any structure (it needs no ellipticity or parabolicity conditions), and the fractional behavior is in time, space, or both.

In a sense, this type of approximation results reveals the true power of fractional equations, independently of the structural "details" of the single equation under consideration, and shows that *space-fractional and time-fractional equations exhibit a variety of solutions which is much richer and more abundant than in the case of classical diffusion*.

Though space- and time-fractional diffusions can be seen as related aspects of nonlocal phenomena, they arise in different contexts and present important structural differences. The paradigmatic example of space-fractional diffusion is embodied by the fractional Laplacian, that is, a fractional root of the classical Laplace operator. This setting often surfaces from stochastic processes presenting jumps and it exhibits the classical spatial symmetries such as invariance under translations and rotations, plus a scale invariance of the integral kernel defining the operator. Differently from this, time-fractional diffusion is typically related to memory effects, and therefore it distinguishes very strongly between the "past" and the "future," and the arrow of time plays a major role (in particular, since the past influences the future, but not viceversa, time-fractional diffusion does not possess the same type of symmetries of the space-fractional one). In these pages, we will be able to consider operators which arise as superpositions of both space- and time-fractional diffusion, possibly taking into account classical derivatives as well (the cases of diffusion which is fractional just in either space or time are comprised as special situations of our general framework). Interestingly, we will also consider fractional operators of any order, showing, in a sense, that some properties related to fractional diffusion persist also when higher order operators come into play, differently from what happens in the classical case, in which the theory available for the Laplacian operator presents significant differences with respect to the case of polyharmonic operators.

To achieve the original result presented here, we develop a broad theory of some fundamental facts about space- and time-fractional equations. Some of these additional results are known from the literature, at least in some particular cases, but some other are new and interesting in themselves, and, in developing these auxiliary theories, this monograph presents a completely self-contained approach to a number of basic questions, such as:

- boundary behavior for the time-fractional eigenfunctions;
- boundary behavior for the time-fractional harmonic functions;
- Green representation formulas;
- existence and regularity for the first eigenfunction of the (possibly higher order) fractional Laplacian;
- boundary asymptotics of the first eigenfunctions of the (possibly higher order) fractional Laplacian;
- boundary behavior of (possibly higher order) fractional harmonic functions.

We now dive into the technical details of this matter.

Acknowledgment

Supported by the Australian Research Council Discovery Project 170104880 NEW "Nonlocal Equations at Work," the DECRA Project DE180100957 "PDEs, free boundaries and applications," and the Fulbright Foundation. The authors are members of INdAM/GNAMPA.

https://doi.org/10.1515/9783110664355-202

Contents

1 Introduction: why fractional derivatives?

The goal of this introductory chapter is to provide a series of simple examples in which fractional diffusion and fractional derivatives naturally arise and give a glimpse on how to use analytical methods to attack simple problems arising in concrete situations. Some of the examples that we present are original, some are modifications of known ones, all will be treated in a *fully elementary* way that requires basically no main prerequisites. Other very interesting motivations can already be found in the specialized literature; see, e. g., [9, 56, 41, 73, 51, 45, 23, 16, 7, 31] (also, in Chapter 2 we will recall some other, somehow more advanced, applications).

Some disclaimers here are mandatory. First of all, the motivations that we provide do not aim at being fully exhaustive, since the number of possible applications of fractional calculus are so abundant that it is virtually impossible to discuss them all in one shot. Moreover (differently from the rest of this monograph), while providing these motivations we do not aim at the maximal mathematical rigor (e. g., all functions will be implicitly assumed to be smooth and suitably decay at infinity, limits will freely be taken and interchanged, etc.), but rather at showing natural contests in which fractional objects appear in an almost unavoidable way.

Example 1.1 (Sliding time and tautochrone problem). Let us consider a point mass subject to gravity, sliding down on a curve without friction. We suppose that the particle starts its motion with zero velocity at height h and it reaches its minimal position at height 0 at time $T(h)$. Our objective is to describe $T(h)$ in terms of the shape of the slide. To this end, see Figure 1.1, we follow an approach introduced by N. H. Abel (see pages 11–27 in [5]) and use coordinates $(x, y) \in \mathbb{R}^2$ to describe the slide as a function of the vertical variable, namely, $x = f(y)$. It is also convenient to introduce the arclength parameter

$$\phi(y) := \sqrt{|f'(y)|^2 + 1} \tag{1.1}$$

and to describe the position of the particle by the notation $p(t) := (f(y(t)), y(t))$. The velocity of the particle is therefore

$$v(t) := \dot{p}(t) = (f'(y(t)), 1)\dot{y}(t). \tag{1.2}$$

By construction, we know that $y(0) = h$ and $y(T(h)) = 0$, and moreover $v(0) = 0$. Accordingly, by the Energy Conservation Principle, for all $t \in [0, T(h)]$,

$$\frac{m|v(t)|^2}{2} + mgy(t) = \frac{m|v(0)|^2}{2} + mgy(0) = mgh,$$

where m is the mass of the particle and g is the gravity acceleration. As a consequence, simplifying m and recalling (1.1) and (1.2) (and that the particle is sliding

https://doi.org/10.1515/9783110664355-001

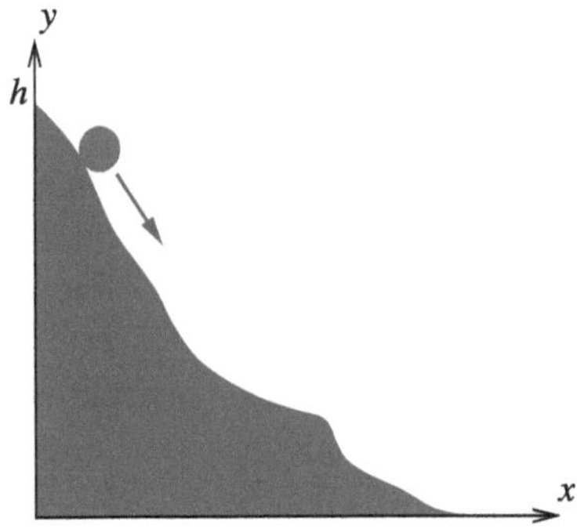

Figure 1.1: A material point sliding down.

downwards),

$$-\phi(y(t))\dot{y}(t) = \sqrt{|f'(y(t))|^2 + 1}\,|\dot{y}(t)| = |v(t)| = \sqrt{2g(h - y(t))}.$$

Hence, separating the variables,

$$T(h) = -\int_0^{T(h)} \frac{\phi(y(t))\dot{y}(t)}{\sqrt{2g(h - y(t))}}\,dt = -\int_h^0 \frac{\phi(y)}{\sqrt{2g(h - y)}}\,dy,$$

that is,

$$T(h) = \int_0^h \frac{\phi(y)}{\sqrt{2g(h - y)}}\,dy. \tag{1.3}$$

We observe that $\phi(y) \geq 1$, thanks to (1.1), and therefore (1.3) gives

$$T(h) \geq \sqrt{\frac{2h}{g}}, \tag{1.4}$$

which corresponds to the free fall, in which f is constant and the particle drops down vertically.

Interestingly, the relation in (1.3) can be seen as a fractional equation. For instance, if we exploit the Caputo fractional derivative of order 1/2, as will be discussed in detail in the forthcoming formula (2.6), one can write (1.3) in the fractional form

$$T(h) = \sqrt{\frac{\pi}{2g}}\,D^{1/2}_{h,0}\Phi(h), \tag{1.5}$$

where Φ is a primitive of ϕ, say,

$$\Phi(H) := \int_0^H \phi(y)\,dy.$$

It is instructive to solve relation (1.3) by obtaining explicitly ϕ in terms of $T(h)$. Of course, fractional calculus, operator algebra, and the theory of Volterra-type integral equations provide general tools to deal with equations such as the one in (1.5), but for the scope of these pages, we try to perform our analysis using only elementary computations. To this end, it is convenient to take advantage of the natural scaling of the problem and convolve (1.3) against the kernel $\frac{1}{\sqrt{h}}$. In this way, we obtain

$$\begin{aligned}\int_0^H \frac{T(h)}{\sqrt{H-h}}\,dh &= \int_0^H \left[\int_0^h \frac{\phi(y)}{\sqrt{2g(h-y)}}\,dy\right]\frac{dh}{\sqrt{H-h}} \\ &= \frac{1}{\sqrt{2g}}\int_0^H \phi(y)\left[\int_y^H \frac{dh}{\sqrt{(h-y)(H-h)}}\right]dy. \end{aligned} \tag{1.6}$$

Using the change of variable $\eta := \frac{h-y}{H-y}$ we see that

$$\int_y^H \frac{dh}{\sqrt{(h-y)(H-h)}} = \int_0^1 \frac{d\eta}{\sqrt{\eta(1-\eta)}} = \pi,$$

and hence (1.6) becomes

$$\int_0^H \frac{T(h)}{\sqrt{H-h}}\,dh = \frac{\pi}{\sqrt{2g}}\int_0^H \phi(y)\,dy = \frac{\pi}{\sqrt{2g}}\Phi(H). \tag{1.7}$$

The main application of this formula consists in a quick solution of the *tautochrone problem*, that is, the determination of the shape of the slide for which the sliding time $T(h)$ is constant (and therefore independent of the initial height h). In this case, we set $T(h) = T$. Then (1.7) gives

$$2T\sqrt{H} = \frac{\pi}{\sqrt{2g}}\Phi(H),$$

and thus, differentiating with respect to H,

$$\frac{T}{\sqrt{H}} = \frac{\pi}{\sqrt{2g}}\phi(H),$$

which, recalling (1.1), leads to

$$|f'(y)|^2 = \phi^2(y) - 1 = \frac{2gT^2}{\pi^2 y} - 1 = \frac{2r-y}{y}, \tag{1.8}$$

with $r := \frac{gT^2}{\pi^2}$, which is the equation of the *cycloid*.

Another interesting application of (1.7) surfaces when the sliding time $T(h)$ behaves like a square root, say,

$$T(h) = \kappa\sqrt{h}, \tag{1.9}$$

for some $\kappa > 0$. In this case, formula (1.7) gives

$$\frac{\kappa\pi H}{2} = \kappa\int_0^H \sqrt{\frac{h}{H-h}}\,dh = \frac{\pi}{\sqrt{2g}}\Phi(H),$$

from which it follows that

$$\phi(H) = \Phi'(H) = \sqrt{\frac{g}{2}}\kappa,$$

and then, by (1.1),

$$|f'(y)|^2 = \phi^2(y) - 1 = \frac{g}{2}\kappa^2 - 1,$$

which is constant (and well-posed when $\kappa \geq \sqrt{\frac{2}{g}}$, coherently with (1.4)). In this case f is linear and therefore the slide corresponds to the classical *inclined plane*.

Example 1.2 (Sliding time and brachistochrone problem). Formula (1.3), as well as its explicit fractional formulation in (1.5), is also useful to solve the *brachistochrone problem*, that is, detecting the curve of fastest descent between two given points. In this setting, the mathematical formulation of Example 1.1 can be modified as follows. The initial height is fixed, hence we can assume that $h > 0$ is a given parameter. Instead, we can optimize the shape of the slide as given by the function f. To this end, it is convenient to write $\phi = \phi_f$ in (1.1), in order to stress its dependence on f. Similarly, the fall time in (1.3) and (1.5) can be denoted by $T(f)$ to emphasize its dependence on f. In this way, we write (1.3) as

$$T(f) = \int_0^h \frac{\phi_f(y)}{\sqrt{2g(h-y)}}\,dy. \tag{1.10}$$

Now, given $\epsilon \in (-1, 1)$, we consider a perturbation $\eta \in C_0^\infty([0, h])$ of an optimal function f. That is, given $\epsilon \in (0, 1)$, we define

$$f_\epsilon(y) := f(y) + \epsilon\eta(y).$$

Since $f_\epsilon(0) = f(0)$ and $f_\epsilon(h) = f(h)$, the endpoints of the slide described by f_ϵ are the same as the ones described by f and therefore the minimality of f gives

$$T(f_\epsilon) = T(f) + o(\epsilon). \tag{1.11}$$

In addition, by (1.1),

$$\begin{aligned}\phi_{f_\epsilon}(y) &= \sqrt{|f'_\epsilon(y)|^2+1} = \sqrt{|f'(y)+\epsilon\eta'(y)|^2+1}\\ &= \sqrt{|f'(y)|^2+2\epsilon f'(y)\eta'(y)+o(\epsilon)+1}\\ &= \sqrt{|f'(y)|^2+1}+\frac{\epsilon f'(y)\eta'(y)}{\sqrt{|f'(y)|^2+1}}+o(\epsilon)\\ &= \phi_f(y)+\frac{\epsilon f'(y)\eta'(y)}{\sqrt{|f'(y)|^2+1}}+o(\epsilon).\end{aligned}$$

Therefore, in light of (1.10) and (1.11),

$$\begin{aligned}o(\epsilon) &= T(f_\epsilon)-T(f)\\ &= \int_0^h \frac{\phi_{f_\epsilon}(y)-\phi_f(y)}{\sqrt{2g(h-y)}}\,dy\\ &= \epsilon\int_0^h \frac{f'(y)\eta'(y)}{\sqrt{2g(h-y)(|f'(y)|^2+1)}}\,dy+o(\epsilon),\end{aligned}$$

and consequently

$$\int_0^h \frac{f'(y)\eta'(y)}{\sqrt{2g(h-y)(|f'(y)|^2+1)}}\,dy = 0.$$

Accordingly, since η is an arbitrary compactly supported perturbation, we obtain the optimality condition

$$\frac{d}{dy}\left(\frac{f'(y)}{\sqrt{2g(h-y)(|f'(y)|^2+1)}}\right)=0$$

and then

$$\frac{f'(y)}{\sqrt{2g(h-y)(|f'(y)|^2+1)}}=-c,$$

for some $c>0$.

This gives

$$|f'(y)|^2=\frac{2c^2g(h-y)}{1-2c^2g(h-y)}. \tag{1.12}$$

It is now expedient to consider a suitable translation of the slide, by considering the rigid motion described by the function

$$\zeta(y) := \frac{y - 1 + 2c^2gh}{2c^2g},$$

and we define

$$\tilde{f}(y) := 2c^2gf(\zeta(y)).$$

We point out that $\zeta'(y) = \frac{1}{2c^2g}$ and

$$\frac{2c^2g(h - \zeta(y))}{1 - 2c^2g(h - \zeta(y))} = \frac{2c^2gh - 2c^2g\zeta(y)}{1 - 2c^2gh + 2c^2g\zeta(y)} = \frac{2c^2gh - (y - 1 + 2c^2gh)}{1 - 2c^2gh + (y - 1 + 2c^2gh)} = \frac{1 - y}{y}.$$

Consequently, by (1.12),

$$|\tilde{f}'(y)|^2 = (2c^2g)^2(\zeta'(y))^2|f'(\zeta(y))|^2 = \frac{2c^2g(h - \zeta(y))}{1 - 2c^2g(h - \zeta(y))} = \frac{1 - y}{y},$$

which is again an equation describing a *cycloid* (compare with (1.8)).

Some additional comments about Examples 1.1 and 1.2 follow. The tautochrone problem was first solved by Christiaan Huygens in his book *Horologium Oscillatorium: sive de motu pendulorum ad horologia aptato demonstrationes geometricae*, published in 1673. Interestingly, the tautochrone problem is also alluded in a famous passage from the 1851 novel *Moby Dick* by Herman Melville.

As for the brachistochrone problem, Galileo was probably the first to take it into account in 1638 in Theorem XXII and Proposition XXXVI of his book *Discorsi e Dimostrazioni Matematiche intorno a due nuove scienze*, in which he seemed to argue that the fastest motion from one end to the other does not take place along the shortest line but along a circular arc (now we know that “circular arc” should have been replaced by “cycloid” to make the statement fully correct, and in fact the name of cycloid is likely to go back to Galileo in its meaning of “resembling a circle”).

The brachistochrone problem was then posed explicitly in 1696 by Johann Bernoulli in *Acta Eruditorum*. Although probably knowing how to solve it himself, Johann Bernoulli challenged all others to solve it, addressing his community with a rather bombastic dare, such as “*I, Johann Bernoulli, address the most brilliant mathematicians in the world. Nothing is more attractive to intelligent people than an honest, challenging problem, whose possible solution will bestow fame and remain as a lasting monument.*”

The coincidence of the solutions of the tautochrone and the brachistochrone problem appears to be a surprising mathematical fact and to reveal deep laws of physics: in the words of Johann Bernoulli, “*Nature always tends to act in the simplest way, and so it here lets one curve serve two different functions.*”

For a detailed historical introduction to these problems, see, e. g., [38] and the references therein.

Example 1.3 (Thermal insulation problems, Dirichlet-to-Neumann problems, and the fractional Laplacian). Let us consider a room whose perimeter is given by walls of two different types. One kind of walls is such that we can fix the temperature there, i. e., by some heaters or coolers endowed with a thermostat. The other kind of walls is made up of insulating material which prevents heat flow to go through. The question that we want to address is:

$$\textit{What is the temperature at the insulating kind of walls?} \tag{1.13}$$

We will see that the question in (1.13) can be conveniently set into a natural fractional Laplacian framework. As a matter of fact, to formalize this question, and address it at least in its simplest possible formulation, let us consider the case in which the room is modeled by the half-space $\mathbb{R}^{n+1}_+ := \mathbb{R}^n \times (0, +\infty)$ (rooms in real life are considered to be three-dimensional, hence n would be equal to 2 in this model; we can also take into account the case of a general n in this discussion). The walls of this room are given by $\mathbb{R}^n \times \{0\}$. We suppose that the insulating material is placed in a nice bounded domain $\Omega \subset \mathbb{R}^n \times \{0\}$ and the temperature is prescribed at the remaining part of the walls $(\mathbb{R}^n \times \{0\}) \setminus \Omega$; see Figure 1.2.

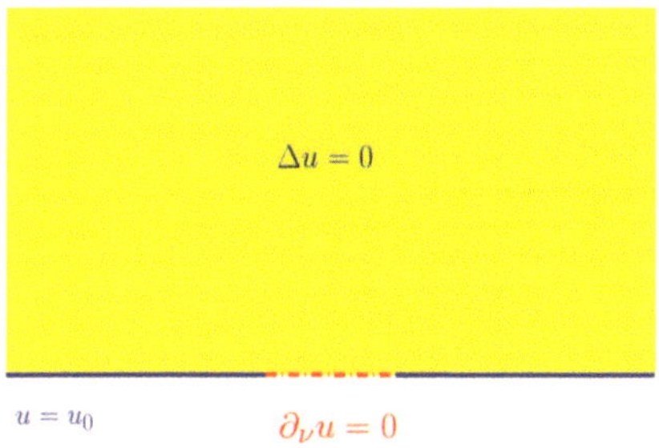

Figure 1.2: The thermal insulation problem in Example 1.3.

The temperature of the room at the point $x = (x_1, \ldots, x_{n+1}) \in \mathbb{R}^n \times [0, +\infty)$ will be described by a function $u = u(x)$. At equilibrium, no heat flow occurs inside the room. Taking the classical ansatz that the heat flow is produced by the gradient of the temperature, since no heat sources are placed inside the room, we obtain that for any ball $B \Subset \mathbb{R}^{n+1}_+$ the heat flux through the boundary of B is necessarily zero and therefore, by the Divergence Theorem,

$$0 = \int_B \operatorname{div}(\nabla u) = \int_B \Delta u,$$

which gives $\Delta u = 0$ in $\mathbb{R}^{n+1}_+$.

Complementing this equation with the prescriptions along the walls, we obtain that the room temperature u satisfies

$$\begin{cases} \Delta u(x) = 0 & \text{for all } x \in \mathbb{R}^{n+1}_+, \\ \partial_{x_{n+1}} u(x) = 0 & \text{for all } x \in \Omega, \\ u(x) = u_0(x_1, \dots, x_n) & \text{for all } x \in (\mathbb{R}^n \times \{0\}) \setminus \Omega. \end{cases} \tag{1.14}$$

As a matter of fact, the setting in (1.14) lacks uniqueness, since if u is a solution of (1.14), then so are

$$u(x) + x_{n+1}, \quad u(x) + x_1 x_{n+1}, \quad u(x) + e^{x_1} \sin x_{n+1},$$

and so on. Hence, to avoid uniqueness issues, we implicitly assume that the solution of (1.14) is constructed by energy minimization: in this way, the strict convexity of the energy functional

$$\int_{\mathbb{R}^{n+1}_+} |\nabla u(x)|^2 \, dx$$

guarantees that the solution is unique.

The question in (1.13) is therefore reduced to find the value of u in Ω. To do so, one can observe that the model links the Neumann and the Dirichlet boundary data of an elliptic problem, namely, given on the boundary the homogeneous Neumann datum in Ω and the (possibly inhomogeneous) Dirichlet datum in $(\mathbb{R}^n \times \{0\}) \setminus \Omega$, one can consider the (minimal energy) harmonic function satisfying these conditions and then calculate its Dirichlet datum in Ω to give an answer to (1.13).

Computationally, it is convenient to observe that equation (1.14) is linear and can therefore efficiently be solved by Fourier transform. Indeed, we write $\hat{u} = \hat{u}(\xi, x_{n+1})$ as the Fourier transform of u in the variables $(x_1, \dots, x_n)$, that is, up to normalizing constants that we neglect for the sake of simplicity, for any $\xi = (\xi_1, \dots, \xi_n) \in \mathbb{R}^n$ and any $x_{n+1} > 0$ we define

$$\hat{u}(\xi, x_{n+1}) := \int_{\mathbb{R}^n} u(x_1, \dots, x_n, x_{n+1}) \exp\left(-i \sum_{j=1}^{n} x_j \xi_j \right) dx_1 \cdots dx_n.$$

Hence, integrating by parts, one sees that, for all $k \in \{1, \dots, n\}$,

$$\begin{aligned} &\int_{\mathbb{R}^n} \partial_{x_k} u(x_1, \dots, x_n, x_{n+1}) \exp\left(-i \sum_{j=1}^{n} x_j \xi_j \right) dx_1 \cdots dx_n \\ &\quad = -i\xi_k \int_{\mathbb{R}^n} u(x_1, \dots, x_n, x_{n+1}) \exp\left(-i \sum_{j=1}^{n} x_j \xi_j \right) dx_1 \cdots dx_n \\ &\quad = -i\xi_k \hat{u}(\xi, x_{n+1}). \end{aligned}$$

Iterating this argument, one obtains, for all $k \in \{1, \dots, n\}$,

$$\int_{\mathbb{R}^n} \partial^2_{x_k} u(x_1, \dots, x_n, x_{n+1}) \exp\left(-i \sum_{j=1}^{n} x_j \xi_j\right) dx_1 \cdots dx_n$$
$$= (-i\xi_k)^2 \hat{u}(\xi, x_{n+1}) = -\xi_k^2 \hat{u}(\xi, x_{n+1}), \tag{1.15}$$

and hence, summing over k and taking into account also the derivatives in x_{n+1},

$$\begin{aligned}
0 &= \widehat{\Delta u}(\xi, x_{n+1}) \\
&= \sum_{k=1}^{n+1} \int_{\mathbb{R}^n} \partial^2_{x_k} u(x_1, \dots, x_n, x_{n+1}) \exp\left(-i \sum_{j=1}^{n} x_j \xi_j\right) dx_1 \cdots dx_n \\
&= -\sum_{k=1}^{n} \xi_k^2 \hat{u}(\xi, x_{n+1}) + \int_{\mathbb{R}^n} \partial^2_{x_{n+1}} u(x_1, \dots, x_n, x_{n+1}) \exp\left(-i \sum_{j=1}^{n} x_j \xi_j\right) dx_1 \cdots dx_n \\
&= -|\xi|^2 \hat{u}(\xi, x_{n+1}) + \partial^2_{x_{n+1}} \hat{u}(\xi, x_{n+1}).
\end{aligned}$$

Consequently, for any $x_{n+1} > 0$,

$$\begin{aligned}
-\partial_{x_{n+1}} \widehat{u(\cdot, 0)} &= -\partial_{x_{n+1}} \hat{u}(\xi, 0) \\
&= -\partial_{x_{n+1}} \hat{u}(\xi, x_{n+1}) + \int_0^{x_{n+1}} \partial^2_{x_{n+1}} \hat{u}(\xi, y)\, dy \\
&= -\partial_{x_{n+1}} \hat{u}(\xi, x_{n+1}) + \int_0^{x_{n+1}} |\xi|^2 \hat{u}(\xi, y)\, dy.
\end{aligned}$$

This equation has solution

$$\hat{u}(\xi, x_{n+1}) = \hat{u}_0(\xi) e^{-|\xi| x_{n+1}},$$

and hence

$$\partial_{x_{n+1}} \hat{u}(\xi, x_{n+1}) = -|\xi| \hat{u}_0(\xi) e^{-|\xi| x_{n+1}} = -|\xi| \hat{u}(\xi, x_{n+1}). \tag{1.16}$$

Combining this with the homogeneous Neumann condition in (1.14), we obtain that, if $\mathcal{F}^{-1}$ denotes the anti-Fourier transform, then

$$\mathcal{F}^{-1}(|\xi| \hat{u}(\cdot, 0)) = 0 \quad \text{in } \Omega. \tag{1.17}$$

It is convenient to write this using the fractional Laplace formulation. As a matter of fact, for every $s \in [0, 1]$ and a (sufficiently smooth and decaying) function $w : \mathbb{R}^n \to \mathbb{R}^n$, one can define

$$(-\Delta)^s w := \mathcal{F}^{-1}(|\xi|^{2s} \hat{w}). \tag{1.18}$$

We observe that when $s = 1$ this definition, up to normalization constants, gives the classical operator $-\Delta$, thanks to (1.15). By a direct computation (see, e. g., Proposition 3.3 in [22]) one also sees that for every $s \in (0,1)$ the operator in (1.18) can be written in integral form as

$$(-\Delta)^s w(x) = \int_{\mathbb{R}^n} \frac{2w(x) - w(x+y) - w(x-y)}{|y|^{n+2s}}\, dy. \tag{1.19}$$

Comparing (1.17) with (1.18), we obtain that equation (1.14) can be written as a fractional equation for a function of n variables (rather than a classical reaction–diffusion equation for a function of $n+1$ variables): indeed, if we set $v(x_1,\dots,x_n) := u(x_1,\dots,x_n,0)$, then

$$\begin{cases} \sqrt{-\Delta} v = 0 & \text{in } \Omega_0, \\ v = u_0 & \text{in } \mathbb{R}^n \setminus \Omega_0, \end{cases} \tag{1.20}$$

where $\Omega_0 \subset \mathbb{R}^n$ is such that $\Omega = \Omega_0 \times \{0\}$.

The solution of the question posed in (1.13) can then be obtained by considering the values of the solution v of (1.20) in Ω_0.

The equivalence between (1.14) and (1.20) (which can also be extended and generalized in many forms) is often quite useful since it allows one to connect classical reaction–diffusion equations and fractional equations and permits the methods typical of one research context to be applicable to the other.

Example 1.4 (The thin obstacle problem). The classical obstacle problem considers an elastic membrane possibly subject to an external force field which is constrained above an obstacle. The elasticity of the membrane makes its graph to be a supersolution in the whole domain and a solution wherever it does not touch the obstacle. For instance, if the vertical force field is denoted by $h : \mathbb{R}^n \to \mathbb{R}$ and the obstacle is given by the subgraph of a function $\varphi : \mathbb{R}^n \to \mathbb{R}$, considering a global problem for the sake of simplicity, the discussion above formalizes in the system of equations

$$\begin{cases} \Delta u \le h & \text{in } \mathbb{R}^n, \\ u \ge \varphi & \text{in } \mathbb{R}^n, \\ \Delta u = 0 & \text{in } \mathbb{R}^n \cap \{u > \varphi\}. \end{cases}$$

As a variation of this problem, one can consider the case in which the obstacle is "thin," i. e., it is supported along a manifold of smaller dimension – concretely, in our case, of codimension 1. For concreteness, one can consider the case in which the obstacle is supported on the hyperplane $\{x_n = 0\}$. In this case, one considers the subgraph in $\{x_n = 0\}$ of a function $\varphi : \mathbb{R}^{n-1} \to \mathbb{R}$ and requires the solution to lie above it; see Figure 1.3.

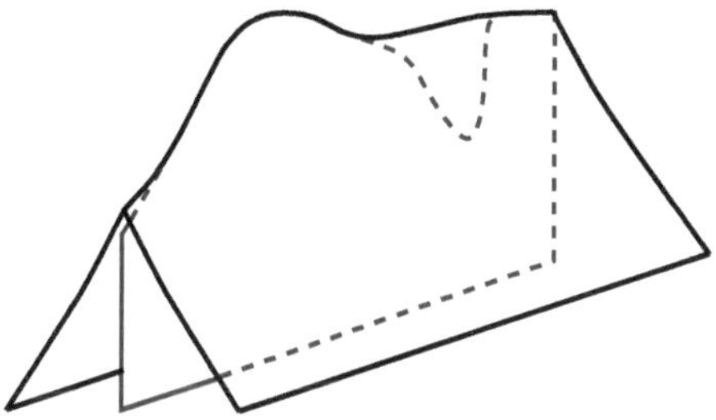

Figure 1.3: The thin obstacle problem.

Combining the thin obstacle constraint with the elasticity of the membrane, we model this problem by the system of equations

$$\begin{cases} \Delta u \leq h & \text{in } \mathbb{R}^n, \\ u \geq \varphi & \text{in } \{x_n = 0\}, \\ \Delta u = 0 & \text{in } \mathbb{R}^n \setminus (\{x_n = 0\} \cap \{u = \varphi\}). \end{cases} \tag{1.21}$$

For concreteness, we will take $h := 0$ from now on (the general case can be reduced to this by subtracting a particular solution). Also, given the structure of (1.21), we will focus on the case of even solutions with respect to the hyperplane $\{x_n = 0\}$, namely,

$$u(x_1, \dots, x_{n-1}, -x_n) = u(x_1, \dots, x_{n-1}, x_n).$$

We observe that, if $\psi \in C_0^\infty(\mathbb{R}^n, [0, +\infty))$, then

$$\begin{aligned} \int_{\mathbb{R}^n} \nabla u(x) \cdot \nabla \psi(x)\, dx &= \int_{\{x_n>0\}} \nabla u(x) \cdot \nabla \psi(x)\, dx + \int_{\{x_n<0\}} \nabla u(x) \cdot \nabla \psi(x)\, dx \\ &= \int_{\{x_n>0\}} \operatorname{div}(\psi(x) \nabla u(x))\, dx + \int_{\{x_n<0\}} \operatorname{div}(\psi(x) \nabla u(x))\, dx \\ &= -2 \int_{\{x_n=0\}} \psi(x) \frac{\partial u}{\partial x_n}(x_1, \dots, x_{n-1}, 0^+)\, d\mathcal{H}^{n-1}(x), \end{aligned}$$

where $\mathcal{H}^{n-1}$ denotes the standard Hausdorff measure of codimension 1. Therefore, the condition $\Delta u \leq 0$ in (1.21) is distributionally equivalent to

$$\frac{\partial u}{\partial x_n}(x_1, \dots, x_{n-1}, 0^+) \leq 0.$$

Similarly, given any point $p \in \{x_n = 0\} \cap \{u > \varphi\}$, one can consider functions $\psi \in C_0^\infty(B_\rho(p))$, with $\rho > 0$ sufficiently small such that $B_\rho(p) \subseteq \{u > \varphi\}$, and thus find that

$$\frac{\partial u}{\partial x_n}(x_1,\dots,x_{n-1},0^+)=0 \quad \text{in } \{x_n=0\}\cap\{u>\varphi\}.$$

Hence, dropping the notation 0^+ for the sake of brevity, one can write (1.21) in the form

$$\begin{cases} \Delta u = 0 & \text{in } \mathbb{R}^n \setminus (\{x_n=0\}\cap\{u=\varphi\}),\\ u\geq\varphi & \text{in } \{x_n=0\},\\ \frac{\partial u}{\partial x_n}\leq 0 & \text{in } \{x_n=0\},\\ \frac{\partial u}{\partial x_n}=0 & \text{in } \{x_n=0\}\cap\{u>\varphi\}. \end{cases} \tag{1.22}$$

Interestingly, equation (1.22) can be written in a fractional Laplacian form. Indeed, by using the Fourier transform in the variables $(x_1,\dots,x_{n-1})$ (see, e. g., (1.16) and (1.18)), we know that, up to normalization constants,

$$\frac{\partial u}{\partial x_n}(x_1,\dots,x_{n-1},0^+) = -\mathcal{F}^{-1}\big(|(\xi_1,\dots,\xi_{n-1})|\hat{u}(\xi_1,\dots,\xi_{n-1},0)\big) = -\sqrt{-\Delta}u(x_1,\dots,x_{n-1},0).$$

Consequently, writing $v := u(x_1,\dots,x_{n-1},0)$, we can interpret (1.22) as a fractional equation on $\mathbb{R}^{n-1}$, namely,

$$\begin{cases} v\geq\varphi & \text{in } \mathbb{R}^{n-1},\\ \sqrt{-\Delta}v\geq 0 & \text{in } \mathbb{R}^{n-1},\\ \sqrt{-\Delta}v=0 & \text{in } \{v>\varphi\}. \end{cases}$$

We do not go into the details of the classical and recent developments of the mathematical theory of the thin obstacle problem and of the many topics related to it. We refer the interested reader to, e. g., [55, 21].

Example 1.5 (The Signorini problem). In 1959, Antonio Signorini posed an engineering problem consisting in finding the equilibrium configuration of an elastic body, resting on a rigid frictionless surface (see [68]). Gaetano Fichera provided a rigorous mathematical framework in which the problem is well-posed in 1963; see [30]. Interestingly, this solution was found just a few weeks before Signorini's death, whose last words were spent to celebrate this discovery as his "*greatest contentment*." The historical description of these moments is commemorated in the survey *La nascita della teoria delle disequazioni variazionali ricordata dopo trent'anni*, of the 1995 *Atti dei Convegni Lincei*.

Here, we recall a simplified version of the problem, its relation with the thin obstacle problem, and its link with the fractional Laplace operator.

To this end, we introduce some notation from linear elasticity. Namely, given a material body $A \subset \mathbb{R}^n$ at rest one describes its equilibrium configuration under suitable forces by $B = y(A)$, where $y : \mathbb{R}^n \to \mathbb{R}^n$ is a (suitably regular and invertible) map (or, equivalently, one can consider B as the set at rest and A as the equilibrium configuration in the "real world," up to replacing y with its inverse).

In this setting, the displacement vector is defined by

$$U(x) := y(x) - x. \tag{1.23}$$

We denote the components of U by $U^{(1)}, \dots, U^{(n)}$. The ansatz of the linear elasticity theory is that, as a consequence of Hooke's Law, the infinitesimal elastic energy is proportional to the symmetrized gradient of U. That is, for every $i, j \in \{1, \dots, n\}$, one defines the strain tensor

$$(\mathcal{D}U)_{ij} := \frac{\partial U^{(i)}}{\partial x_j} + \frac{\partial U^{(j)}}{\partial x_i} \tag{1.24}$$

and sets

$$|\mathcal{D}U| := \sum_{i,j=1}^{n} |(\mathcal{D}U)_{ij}|^2 = \sum_{i,j=1}^{n} \left(\frac{\partial U^{(i)}}{\partial x_j} + \frac{\partial U^{(j)}}{\partial x_i} \right)^2.$$

With this notation, the elastic component of the energy is

$$\mathcal{E}(U) := \frac{1}{2} \int_A |\mathcal{D}U(x)|^2 \, dx. \tag{1.25}$$

The differential operator governing the elastostatic equations (known in the literature as Navier–Cauchy equations, at least in the special form that we take into account in our simplified approach) are obtained from the first variation of the elastic energy functional in (1.25). Since

$$\begin{aligned}
&\frac{|\mathcal{D}(U + \epsilon\Phi)|^2 - |\mathcal{D}U|^2}{2} \\
&= \frac{1}{2} \sum_{i,j=1}^{n} \left(\frac{\partial U^{(i)}}{\partial x_j} + \frac{\partial U^{(j)}}{\partial x_i} + \epsilon \left(\frac{\partial \Phi^{(i)}}{\partial x_j} + \frac{\partial \Phi^{(j)}}{\partial x_i} \right) \right)^2 - \frac{|\mathcal{D}U|^2}{2} \\
&= \epsilon \sum_{i,j=1}^{n} \left(\frac{\partial U^{(i)}}{\partial x_j} + \frac{\partial U^{(j)}}{\partial x_i} \right) \left(\frac{\partial \Phi^{(i)}}{\partial x_j} + \frac{\partial \Phi^{(j)}}{\partial x_i} \right) + o(\epsilon) \\
&= \epsilon \sum_{i,j=1}^{n} \left(\frac{\partial U^{(i)}}{\partial x_j} \frac{\partial \Phi^{(i)}}{\partial x_j} + \frac{\partial U^{(j)}}{\partial x_i} \frac{\partial \Phi^{(i)}}{\partial x_j} + \frac{\partial U^{(i)}}{\partial x_j} \frac{\partial \Phi^{(j)}}{\partial x_i} + \frac{\partial U^{(j)}}{\partial x_i} \frac{\partial \Phi^{(j)}}{\partial x_i} \right) + o(\epsilon) \\
&= \epsilon \sum_{i,j=1}^{n} \left(\frac{\partial U^{(i)}}{\partial x_j} \frac{\partial \Phi^{(i)}}{\partial x_j} + \frac{\partial U^{(j)}}{\partial x_i} \frac{\partial \Phi^{(i)}}{\partial x_j} + \frac{\partial U^{(j)}}{\partial x_i} \frac{\partial \Phi^{(i)}}{\partial x_j} + \frac{\partial U^{(i)}}{\partial x_j} \frac{\partial \Phi^{(i)}}{\partial x_j} \right) + o(\epsilon) \\
&= 2\epsilon \sum_{i,j=1}^{n} \left(\frac{\partial U^{(i)}}{\partial x_j} + \frac{\partial U^{(j)}}{\partial x_i} \right) \frac{\partial \Phi^{(i)}}{\partial x_j} + o(\epsilon),
\end{aligned}$$

it follows from (1.25) that, for all $\Phi \in C_0^\infty(A, \mathbb{R}^n)$,

$$
\begin{aligned}
\langle D\mathcal{E}(U), \Phi \rangle &= 2 \sum_{i,j=1}^{n} \int_A \left(\frac{\partial U^{(i)}}{\partial x_j}(x) + \frac{\partial U^{(j)}}{\partial x_i}(x) \right) \frac{\partial \Phi^{(i)}}{\partial x_j}(x)\, dx \\
&= -2 \sum_{i,j=1}^{n} \int_A \frac{\partial}{\partial x_j} \left(\frac{\partial U^{(i)}}{\partial x_j}(x) + \frac{\partial U^{(j)}}{\partial x_i}(x) \right) \Phi^{(i)}(x)\, dx \\
&= -2 \sum_{i=1}^{n} \int_A \left(\Delta U^{(i)}(x) + \frac{\partial}{\partial x_i} \operatorname{div} U(x) \right) \Phi^{(i)}(x)\, dx. \qquad (1.26)
\end{aligned}
$$

One can also take into account the effect of a force field $f = (f^{(1)}, \dots, f^{(n)}) : \mathbb{R}^n \to \mathbb{R}^n$ which acts on the material body. If we think that the map y is the outcome of the deformation produced by the force field, one can consider the infinitesimal work associated to this force as given approximatively, for small displacements, by the quantity

$$f(x) \cdot (y(x) - x)\, dx = f(x) \cdot U(x)\, dx,$$

where the setting in (1.23) has been exploited. We obtain in this way a potential energy of the form

$$\mathcal{P}(U) := \int_A f(x) \cdot U(x)\, dx.$$

We see that

$$\langle \nabla \mathcal{P}(U), \Phi \rangle = \int_A f(x) \cdot \Phi(x)\, dx, \qquad (1.27)$$

for all $\Phi \in C_0^\infty(A, \mathbb{R}^n)$.

From this and (1.26), we obtain the elastic equation

$$
\begin{gathered}
\Delta U^{(i)}(x) + \frac{\partial}{\partial x_i} \operatorname{div} U(x) = \frac{1}{2} f^{(i)}(x), \\
\text{for all } i \in \{1, \dots, n\} \text{ and all } x \in A. \qquad (1.28)
\end{gathered}
$$

We also assume that $y(\partial A) = \partial B$ and that B rests on a rigid surface, say, a frictionless table. In this case, we can write $A \subset \{x_n \geq 0\}$ and $B \subset \{y_n \geq 0\}$, that is, $y^{(n)}(x) > 0$ for all $x \in A$, and therefore $U^{(n)}(x) + x_n > 0$ for all $x \in A$.

The contact set between B and the table is described by the points lying in $T := (\partial B) \cap \{y_n = 0\}$, and the contact set between A and the table is described by the points lying in $S := (\partial A) \cap \{x_n = 0\}$. We describe S by distinguishing two classes of points, namely, the points S_1 which do not leave the contact set and the points S_2 which leave the contact set, namely,

$$\begin{aligned} S_1 &:= \{x \in S \text{ s.t. } y^{(n)}(x) = 0\} \\ &= \{x \in S \text{ s.t. } U^{(n)}(x) + x_n = 0\} \\ &= \{x \in S \text{ s.t. } U^{(n)}(x) = 0\} \\ \text{and} \quad S_2 &:= \{x \in S \text{ s.t. } y^{(n)}(x) > 0\} \\ &= \{x \in S \text{ s.t. } U^{(n)}(x) + x_n > 0\} \\ &= \{x \in S \text{ s.t. } U^{(n)}(x) > 0\}. \end{aligned} \tag{1.29}$$

We take into account the effect of a surface tension, acting by means of a force field $g : \mathbb{R}^n \to \mathbb{R}^n$. In this case, the infinitesimal work, for small displacements, is approximately given by

$$g(x) \cdot (y(x) - x)\, d\mathcal{H}^{n-1}(x) = g(x) \cdot U(x)\, d\mathcal{H}^{n-1}(x).$$

Therefore, we describe the surface tension energy effect by an energy functional of the form

$$\mathcal{S}(U) := \int_{\partial A} g(x) \cdot U(x)\, d\mathcal{H}^{n-1}(x).$$

We stress that if $x_0 \in S_2$ there exist $\rho > 0$ and $\epsilon_0 > 0$ such that for all $\epsilon \in [-\epsilon_0, \epsilon_0]$ and $\Phi \in C_0^\infty(B_\rho(x_0), \mathbb{R}^n)$ the perturbation $U + \epsilon\Phi$ is admissible, in the sense that it maps A into $y(A) \subset \{y_n \geq 0\}$.

On the other hand, if $\Phi \in C_0^\infty(\mathbb{R}^n, \mathbb{R}^n)$ and $\Phi^{(n)} \geq 0$ the perturbation $U + \epsilon\Phi$ is admissible for all $\epsilon \geq 0$, since, for every $x \in A$,

$$0 \leq y^{(n)}(x) = U^{(n)}(x) + x_n \leq U^{(n)}(x) + \epsilon\Phi^{(n)}(x) + x_n.$$

Furthermore,

$$\langle D\mathcal{S}(U), \Phi\rangle = \int_{\partial A} g(x) \cdot \Phi(x)\, d\mathcal{H}^{n-1}(x),$$

for all $\Phi \in C_0^\infty(\mathbb{R}^n, \mathbb{R}^n)$. Therefore, recalling (1.26), (1.27), and (1.28), it follows that the variation of the full energy with respect to a boundary perturbation $\Phi \in C_0^\infty(\mathbb{R}^n, \mathbb{R}^n)$ is equal to

$$\begin{aligned} &2\sum_{i,j=1}^{n} \int_A \left(\frac{\partial U^{(i)}}{\partial x_j}(x) + \frac{\partial U^{(j)}}{\partial x_i}(x)\right)\frac{\partial \Phi^{(i)}}{\partial x_j}(x)\, dx \\ &\quad + \int_A f(x) \cdot \Phi(x)\, dx + \int_{\partial A} g(x) \cdot \Phi(x)\, d\mathcal{H}^{n-1}(x) \end{aligned}$$

$$
\begin{aligned}
&= 2\sum_{i,j=1}^{n}\int_A \frac{\partial}{\partial x_j}\left(\left(\frac{\partial U^{(i)}}{\partial x_j}(x) + \frac{\partial U^{(j)}}{\partial x_i}(x)\right)\Phi^{(i)}(x)\right)dx \\
&\quad + \int_{\partial A} g(x)\cdot\Phi(x)\,d\mathcal{H}^{n-1}(x) \\
&= 2\sum_{i=1}^{n}\int_A \operatorname{div}\left(\Phi^{(i)}(x)\left(\nabla U^{(i)}(x) + \frac{\partial U}{\partial x_i}(x)\right)\right)dx \\
&\quad + \int_{\partial A} g(x)\cdot\Phi(x)\,d\mathcal{H}^{n-1}(x) \\
&= 2\sum_{i=1}^{n}\int_{\partial A} \Phi^{(i)}(x)\left(\nabla U^{(i)}(x) + \frac{\partial U}{\partial x_i}(x)\right)\cdot\nu(x)\,d\mathcal{H}^{n-1}(x) \\
&\quad + \sum_{i=1}^{n}\int_{\partial A} g^{(i)}(x)\Phi^{(i)}(x)\,d\mathcal{H}^{n-1}(x) \\
&= 2\sum_{i,k=1}^{n}\int_{\partial A} \Phi^{(i)}(x)(\mathcal{D}U)_{ik}(x)\nu_k(x)\,d\mathcal{H}^{n-1}(x) \\
&\quad + \sum_{i=1}^{n}\int_{\partial A} g^{(i)}(x)\Phi^{(i)}(x)\,d\mathcal{H}^{n-1}(x),
\end{aligned}
$$

where ν is the exterior normal of A. This and the admissibility discussion of the perturbation gives the boundary conditions

$$
\begin{aligned}
&\sum_{k=1}^{n}(\mathcal{D}U)_{nk}(x)\nu_k(x) = -\frac{1}{2}g^{(n)}(x) \quad \text{for all } x \in S_2 \\
\text{and}\quad &\sum_{k=1}^{n}(\mathcal{D}U)_{nk}(x)\nu_k(x) \geq -\frac{1}{2}g^{(n)}(x) \quad \text{for all } x \in S_1.
\end{aligned} \tag{1.30}
$$

If the surface forces are tangential to the boundary of the material body, $g^{(n)} = 0$ on the surface of the table, and the normal at these points is vertical, so therefore (1.30) reduces to

$$
\begin{aligned}
&(\mathcal{D}U)_{nn}(x) = 0 \quad \text{for all } x \in S_2 \\
\text{and}\quad &(\mathcal{D}U)_{nn}(x) \leq 0 \quad \text{for all } x \in S_1.
\end{aligned} \tag{1.31}
$$

Interestingly, the problem is naturally endowed with “ambiguous” boundary conditions, in the sense that there are two alternative boundary conditions in (1.31) that are prescribed at the boundary, in terms of either equalities and inequalities, and it is not a priori known what condition is satisfied at each point. Also, by (1.24), we can write (1.31) in the form

$$\frac{\partial U^{(n)}}{\partial x_n}(x) = 0 \quad \text{for all } x \in S_2$$
$$\text{and} \quad \frac{\partial U^{(n)}}{\partial x_n}(x) \leq 0 \quad \text{for all } x \in S_1. \tag{1.32}$$

We now consider a magnified version of this picture at "nice" contact points. For the sake of simplicity we suppose that

$$0 \in \partial A, \quad B_\rho(\rho e_n) \subseteq A, \quad B_\rho(-\rho e_1) \cap \{x_n = 0\} \subseteq S_1,$$
$$\text{and} \quad B_\rho(\rho e_1) \cap \{x_n = 0\} \subseteq S_2, \tag{1.33}$$

for some $\rho > 0$; see Figure 1.4.

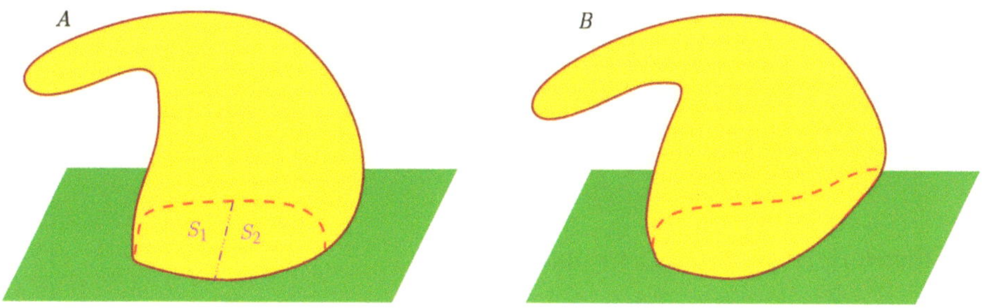

Figure 1.4: Points leaving the contact set.

For fixed $\epsilon > 0$, to be taken small in what follows, we consider the transformation

$$\Theta_\epsilon(x) := \left(\frac{x_1}{\epsilon}, \dots, \frac{x_{n-1}}{\epsilon}, \frac{\sqrt{2}x_n}{\epsilon} \right).$$

By (1.33), we have, locally,

$$\Theta_\epsilon(A) \text{ converges to } \{x_n > 0\},$$
$$\Theta_\epsilon(S_1) \text{ converges to } \{x_n = 0\} \cap \{x_1 < 0\},$$
$$\text{and} \quad \Theta_\epsilon(S_2) \text{ converges to } \{x_n = 0\} \cap \{x_1 > 0\}, \tag{1.34}$$

up to negligible sets.

Moreover, for each $i \in \{1, \dots, n\}$, we define

$$V_\epsilon^{(i)}(x) := \epsilon^2 \left[U^{(i)}(\Theta_\epsilon(x)) - U^{(i)}(0) - \sum_{k=1}^{n} \frac{\partial U^{(i)}}{\partial x_k}(0) \Theta_\epsilon^{(k)}(x) \right]. \tag{1.35}$$

Under a quadratic bound on U, one can make the ansatz that $V_\epsilon^{(i)}$ behaves well as $\epsilon \searrow 0$, and we will indeed assume that $V_\epsilon = (V_\epsilon^{(1)}, \dots, V_\epsilon^{(n)})$ converges smoothly to some $V = (V^{(1)}, \dots, V^{(n)})$. We also set $v := V^{(n)}$. We point out that, by (1.32) and (1.33), we can take sequences of points $p_j \in S_1$ and $q_j \in S_2$ which converge to the origin as $j \to +\infty$, and, assuming that the solution is regular enough, we can write

$$\begin{aligned} & 0 = \lim_{j\to+\infty} U^{(n)}(p_j) = U^{(n)}(0) \\ \text{and} \quad & 0 = \lim_{j\to+\infty} \frac{\partial U^{(n)}}{\partial x_n}(q_j) = \frac{\partial U^{(n)}}{\partial x_n}(0). \end{aligned} \tag{1.36}$$

As a consequence, if $\Theta_\epsilon(x) \in S_1 \cup S_2$,

$$\begin{aligned} V_\epsilon^{(n)}(x) &= \epsilon^2 \left[U^{(n)}(\Theta_\epsilon(x)) - \sum_{k=1}^{n-1} \frac{\partial U^{(i)}}{\partial x_k}(0)\Theta_\epsilon^{(k)}(x) \right] \\ &\geq -\epsilon^2 \sum_{k=1}^{n-1} \frac{\partial U^{(i)}}{\partial x_k}(0)\Theta_\epsilon^{(k)}(x) \\ &= -\epsilon \sum_{k=1}^{n-1} \frac{\partial U^{(i)}}{\partial x_k}(0) x_k, \end{aligned}$$

and therefore, recalling (1.34), we obtain

$$v(x_1, \dots, x_{n-1}, 0) \geq 0, \quad \text{for all } (x_1, \dots, x_{n-1}) \in \mathbb{R}^{n-1}. \tag{1.37}$$

By (1.36), it also follows that

$$\begin{aligned} \frac{\partial V_\epsilon^{(n)}}{\partial x_n}(x) &= \sqrt{2}\epsilon \left[\frac{\partial U^{(n)}}{\partial x_n}(\Theta_\epsilon(x)) - \frac{\partial U^{(n)}}{\partial x_n}(0) \right] \\ &= \sqrt{2}\epsilon \frac{\partial U^{(n)}}{\partial x_n}(\Theta_\epsilon(x)). \end{aligned}$$

Hence, by (1.32),

$$\begin{aligned} & \frac{\partial V_\epsilon^{(n)}}{\partial x_n}(x) = 0 \quad \text{if } \Theta_\epsilon(x) \in S_2 \\ \text{and} \quad & \frac{\partial V_\epsilon^{(n)}}{\partial x_n}(x) \leq 0 \quad \text{if } \Theta_\epsilon(x) \in S_1. \end{aligned} \tag{1.38}$$

This and (1.34) formally lead to

$$\begin{aligned} & \frac{\partial v}{\partial x_n}(x) = 0 \quad \text{in } \{x_n = 0\} \cap \{x_1 > 0\} \\ \text{and} \quad & \frac{\partial v}{\partial x_n}(x) \leq 0 \quad \text{in } \{x_n = 0\} \cap \{x_1 < 0\}. \end{aligned} \tag{1.39}$$

One can make (1.39) more precise in light of (1.37). Namely, we claim that

$$\frac{\partial v}{\partial x_n}(x) = 0 \quad \text{in } \{x_n = 0\} \cap \{v > 0\}. \tag{1.40}$$

To check this, let us take $x \in \{x_n = 0\}$ with $a := v(x) > 0$ and let $\epsilon > 0$ be so small that $V_\epsilon^{(n)}(x) \geq \frac{a}{2}$ and $|\frac{\partial U^{(i)}}{\partial x_k}(0)||x| \leq \frac{a}{4\epsilon(n-1)}$, for all $k \in \{1, \dots, n-1\}$. Then, by (1.35) and (1.36), we have

$$\epsilon^2 U^{(n)}(\Theta_\epsilon(x)) = V_\epsilon^{(n)}(x) + \epsilon^2 \sum_{k=1}^{n-1} \frac{\partial U^{(i)}}{\partial x_k}(0)\Theta_\epsilon^{(k)}(x) \geq \frac{a}{2} - \frac{a}{4} > 0.$$

From this and (1.29) it follows that $\Theta_\epsilon(x) \in S_2$ and consequently (1.40) follows from (1.38).

We also introduce the notation

$$\sigma_k := \begin{cases} 1 & \text{if } k \neq n, \\ \sqrt{2} & \text{if } k = n. \end{cases}$$

Then, we see that, for each $m \in \{1, \dots, n\}$,

$$\frac{\partial^2 V_\epsilon^{(n)}}{\partial x_m^2}(x) = \sigma_m^2 \frac{\partial^2 U^{(n)}}{\partial x_m^2}(\Theta_\epsilon(x)),$$

and accordingly, for every $x \in \mathbb{R}^{n-1} \times (0, +\infty)$ such that $\Theta_\epsilon(x) \in A$,

$$\begin{aligned} \Delta V_\epsilon^{(n)}(x) &= \sum_{m=1}^{n} \sigma_m^2 \frac{\partial^2 U^{(n)}}{\partial x_m^2}(\Theta_\epsilon(x)) \\ &= \sum_{m=1}^{n-1} \frac{\partial^2 U^{(n)}}{\partial x_m^2}(\Theta_\epsilon(x)) + 2\frac{\partial^2 U^{(n)}}{\partial x_n^2}(\Theta_\epsilon(x)). \end{aligned}$$

Hence, since, by (1.28),

$$\begin{aligned} \frac{1}{2}f^{(n)}(x) &= \Delta U^{(n)}(x) + \frac{\partial}{\partial x_n}\operatorname{div} U(x) \\ &= \sum_{m=1}^{n} \frac{\partial^2 U^{(n)}}{\partial x_m^2}(x) + \sum_{m=1}^{n} \frac{\partial^2 U^{(m)}}{\partial x_m \partial x_n}(x) \\ &= \sum_{m=1}^{n-1} \frac{\partial^2 U^{(n)}}{\partial x_m^2}(x) + \sum_{m=1}^{n-1} \frac{\partial^2 U^{(m)}}{\partial x_m \partial x_n}(x) + 2\frac{\partial^2 U^{(n)}}{\partial x_n^2}(x), \end{aligned}$$

and we find that, if $\Theta_\epsilon(x) \in A$,

$$\frac{1}{2}f^{(n)}(\Theta_\epsilon(x)) = \Delta V_\epsilon^{(n)}(x) + \sum_{m=1}^{n-1} \frac{\partial^2 U^{(m)}}{\partial x_m \partial x_n}(\Theta_\epsilon(x)).$$

In particular, if the force field is due to vertical gravity, we have $f^{(n)} = -g$ for some constant g, and accordingly

$$\Delta V_\epsilon^{(n)}(x) = h(x),$$

as long as $\Theta_\epsilon(x) \in A$, with

$$h(x) := -\frac{g}{2} - \sum_{m=1}^{n-1} \frac{\partial^2 U^{(m)}}{\partial x_m \partial x_n}(\Theta_\epsilon(x)).$$

Hence, by (1.34), we can write

$$\Delta v(x) = h(x),$$

for all $x \in \mathbb{R}^{n-1} \times (0, +\infty)$. Combining this with (1.37), (1.39), and (1.40), we can write the system of equations

$$\begin{cases} \Delta v(x) = h(x) & \text{in } \{x_n > 0\}, \\ v \geq 0 & \text{on } \{x_n = 0\}, \\ \frac{\partial v}{\partial x_n}(x) \leq 0 & \text{on } \{x_n = 0\}, \\ \frac{\partial v}{\partial x_n}(x) = 0 & \text{on } \{x_n = 0\} \cap \{v > 0\}. \end{cases} \tag{1.41}$$

By taking even reflection, one can also define

$$u(x) = u(x_1, \dots, x_n) = \begin{cases} v(x_1, \dots, x_{n-1}, x_n) & \text{if } x_n \geq 0, \\ v(x_1, \dots, x_{n-1}, -x_n) & \text{if } x_n < 0. \end{cases}$$

Similarly, we define h in $\{x_n < 0\}$ by even reflection, and in this way

$$\Delta u(x) = \Delta v(x_1, \dots, x_{n-1}, |x_n|) = h(x_1, \dots, x_{n-1}, |x_n|) = h(x),$$

as long as $x_n \neq 0$.

We also observe that if $\varphi \in C_0^\infty(\mathbb{R}^n)$ is such that $\varphi = 0$ in $\{x_n = 0\} \cap \{v = 0\}$,

$$\begin{aligned} &\int_{\mathbb{R}^n} \nabla u(x) \cdot \nabla \varphi(x)\, dx + \int_{\mathbb{R}^n} h(x)\varphi(x)\, dx \\ &= \int_{\mathbb{R}^n \cap \{x_n > 0\}} \nabla u(x) \cdot \nabla \varphi(x)\, dx + \int_{\mathbb{R}^n \cap \{x_n < 0\}} \nabla u(x) \cdot \nabla \varphi(x)\, dx \\ &\quad + \int_{\mathbb{R}^n \cap \{x_n > 0\}} h(x)\varphi(x)\, dx + \int_{\mathbb{R}^n \cap \{x_n < 0\}} h(x)\varphi(x)\, dx \end{aligned}$$

$$
\begin{aligned}
&= \int\limits_{\mathbb{R}^n\cap\{x_n>0\}} \operatorname{div}(\varphi(x)\nabla u(x))\,dx + \int\limits_{\mathbb{R}^n\cap\{x_n<0\}} \operatorname{div}(\varphi(x)\nabla u(x))\,dx \\
&= -\int\limits_{\{x_n=0\}} \varphi(x)\frac{\partial u}{\partial x_n}(x_1,\dots,x_{n-1},0^+)\,d\mathcal{H}^{n-1}(x) \\
&\quad + \int\limits_{\{x_n=0\}} \varphi(x)\frac{\partial u}{\partial x_n}(x_1,\dots,x_{n-1},0^-)\,d\mathcal{H}^{n-1}(x) \\
&= -2 \int\limits_{\{x_n=0\}\cap\{v>0\}} \varphi(x)\frac{\partial v}{\partial x_n}(x_1,\dots,x_{n-1},0^+)\,d\mathcal{H}^{n-1}(x) \\
&= 0,
\end{aligned}
$$

thanks to (1.41). It follows that

$$\Delta u = h \quad \text{in } \mathbb{R}^n \setminus (\{x_n = 0\} \cap \{u = 0\}).$$

This and (1.41) lead to the system

$$
\begin{cases}
\Delta u(x) = h(x) & \text{in } \mathbb{R}^n \setminus (\{x_n = 0\} \cap \{u = 0\}), \\
u \geq 0 & \text{on } \{x_n = 0\}, \\
\frac{\partial u}{\partial x_n}(x) \leq 0 & \text{on } \{x_n = 0\}, \\
\frac{\partial u}{\partial x_n}(x) = 0 & \text{on } \{x_n = 0\} \cap \{u > 0\},
\end{cases}
$$

which is in the form of the thin obstacle problem[1] discussed in Example 1.4 (compare with (1.22)).

Example 1.6 (**Gamma function, Balakrishnan formula, the method of semigroups, and the Heaviside operational calculus**). Let us start with the classical definition of the Euler Gamma function, i. e.:

$$\Gamma(z) = \int_0^{+\infty} \tau^{z-1} e^{-\tau}\,d\tau.$$

We compute it at the point $z := 1 - s$, with $s \in (0,1)$, and we integrate by parts, thus obtaining

1 As a matter of fact, in the recent mathematical jargon, there is some linguistic confusion about the "Signorini problem," since this name is often used also for the thin obstacle problem in Example 1.4. In a sense, the thin obstacle problem should be properly referred to as the "scalar" Signorini problem, but the adjective "scalar" often happens to be missing. Of course, the "original" Signorini problem is technically even more demanding than the thin obstacle problem, due to the vectorial nature of the question. For the optimal regularity and the free boundary analysis of the original Signorini problem and some important links with the scalar Signorini problem, we refer to [8] (see in particular Section 5 there).

$$\Gamma(1-s) = \int_0^{+\infty} \tau^{-s} e^{-\tau}\, d\tau$$
$$= -\int_0^{+\infty} \tau^{-s} \frac{d}{d\tau}(e^{-\tau} - 1)\, d\tau$$
$$= -s \int_0^{+\infty} \tau^{-s-1}(e^{-\tau} - 1)\, d\tau,$$

which can be written as

$$\Gamma(-s) = \int_0^{+\infty} \tau^{-s-1}(e^{-\tau} - 1)\, d\tau.$$

Now, we take $\delta > 0$ and make the substitution $t := \delta^{-1}\tau$. In this way, we obtain

$$\delta^s = \frac{1}{\Gamma(-s)} \int_0^{+\infty} t^{-s-1}(e^{-t\delta} - 1)\, dt. \tag{1.42}$$

It turns out that one can make sense of this formula not only for a given real parameter $\delta > 0$, but also when δ is replaced by a suitably nice operator such as the Laplacian (with the minus sign to make it positive). Namely, formally taking $\delta := -\Delta$ in (1.42), one finds that

$$(-\Delta)^s = \frac{1}{\Gamma(-s)} \int_0^{+\infty} t^{-s-1}(e^{t\Delta} - 1)\, dt. \tag{1.43}$$

In spite of the sloppy way in which formula (1.43) was derived here, it is possible to give a rigorous proof of it using operator theory. Indeed, formula (1.43) was established by Alampallam V. Balakrishnan in [10]. Its meaning in the operator sense is that applying the operator on the left hand side of (1.43) to a nice (say, smooth and rapidly decreasing) function is equivalent to applying to it the operator on the right hand side, namely,

$$(-\Delta)^s u(x) = \frac{1}{\Gamma(-s)} \int_0^{+\infty} t^{-s-1}(e^{t\Delta}u(x) - u(x))\, dt. \tag{1.44}$$

The meaning of $e^{t\Delta}u(x)$ is also in the sense of operators. To understand this notation one can set $\Phi_u(x,t) := e^{t\Delta}u(x)$ and observe that, formally,

$$\partial_t \Phi_u(x,t) = \partial_t(e^{t\Delta}u(x)) = \Delta e^{t\Delta}u(x) = \Delta\Phi_u(x,t)$$
$$\text{and} \quad \Phi_u(x,0) = e^{\Delta 0}u(x) = e^0 u(x) = u(x).$$

These observations can be formalized by "going backwards" in the computation and defining $e^{t\Delta}u(x) := \Phi_u(x,t)$, where the latter is the solution of the heat equation with initial datum u, that is,

$$\begin{cases} \partial_t \Phi_u(x,t) = \Delta \Phi_u(x,t) & \text{for all } x \in \mathbb{R}^n \text{ and } t > 0, \\ \Phi_u(x,0) = u(x) & \text{for all } x \in \mathbb{R}^n. \end{cases}$$

With this notation, we can write (1.44) as

$$(-\Delta)^s u(x) = \frac{1}{\Gamma(-s)} \int_0^{+\infty} t^{-s-1}(\Phi_u(x,t) - u(x))\, dt. \tag{1.45}$$

The power of formula (1.45) is apparent, since it reduces a nonlocal, and in principle rather complicated, operator such as the fractional Laplacian to the superposition of classical heat semigroups, and can be exploited as a "subordination identity" in which the well-established knowledge of the classical heat flow leads to new results for the fractional setting; see, e. g., [43]. The importance and broad range of applicability of formula (1.45) and of its various extensions are very clearly and extensively discussed in [31, 71].

Now we make some historical comments about operator calculus and its successful attempt to transform identities valid for real numbers into rigorous formulas involving operators, under the appropriate assumptions. Without aiming at reconstructing here the full history of the subject, we recall that one of the first attempts in the important directions sketched here was made at the end of the nineteenth century by Oliver Heaviside, who also tried to write the solution of the heat equation in an operator form. Given the difficulty of the arguments treated and the lack of mathematical technologies at that time, some of the original arguments in the literature were probably not fully justified and required the introduction of a brand new subject of mathematical analysis, which indeed highly contributed to its creation and promotion; see, e. g., [44]. It is however plausible that the pioneering, albeit somewhat unrigorous, intuitions of Heaviside were not always well appreciated by the mathematical community at that time. A footprint of this historical controversy has remained in the work by Heaviside published in Volume 34 of the periodical and scientific journal *The Electrician*, in which Heaviside states that "*What one has a right to expect, however, is a fair field, and that the want of sympathy should be kept in a neutral state, so as not to lead to unnecessary obstruction. For even men who are not Cambridge mathematicians deserve justice, which I very much fear they do not always get, especially the meek and lowly.*"

For an extensive treatment of the theory of operator calculus, see, e. g., [50, 52, 37, 35] and the references therein.

Example 1.7 (Fractional viscoelastic models, springs, and dashpots). A classical application of fractional derivatives occurs in the phenomenological description of viscoelastic fluids. This is of course a very advanced topic and we do not aim to fully cover it in these few pages; see, e. g., Section 10.2 of [56], where a number of fractional models for viscoelasticity are discussed in detail.

Roughly speaking, a basic idea used in this context is that the viscoelastic effects arise as a suitable "ideal" superposition of "purely elastic" and "purely viscous" phenomena, which are better understood when treated separately but whose combined effect becomes quite difficult to comprise into classical equations.

On the one hand, the elastic effects are well described by the displacement of a classical spring subject to Hooke's Law, in which the displacement of the spring is proportional to the force applied to it. That is, if ε denotes the elongation of the spring and σ the force applied to it, one writes

$$\sigma = \kappa\varepsilon, \tag{1.46}$$

for a suitable elastic coefficient $\kappa > 0$.

On the other hand, the viscous effects of fluids is classically described by Newton's Law, according to which forces are related to velocities, as in the formula

$$\sigma = \nu\dot{\varepsilon}, \tag{1.47}$$

where the "dot" denotes the derivative with respect to time, for a suitable viscous coefficient $\nu > 0$.

The rationale sustaining (1.47) can be understood thinking about the free fall of an object. In this case, if one takes into account the gravity and the viscous friction of the air, the vertical position of a falling object is described by the equation

$$mg - \nu\dot{\varepsilon} = m\ddot{\varepsilon}. \tag{1.48}$$

For long times, the falling body reaches asymptotically a terminal velocity, which can be guessed by (1.48) by formally imposing that the limit acceleration is zero: in this limit regime, one thus obtains the velocity equation

$$mg - \nu\dot{\varepsilon} = 0, \tag{1.49}$$

which formally coincides with equation (1.47) when the force is the gravitational one. Hence, in a sense, comparing (1.47) with (1.49), one can think that Newton's Law for viscid fluids describes the asymptotic velocity in a regime in which the viscous effects are dominant and after a sufficient amount of time (after which the acceleration effects become negligible).

Roughly speaking, the idea of viscoelasticity is that, in general, fluids are neither perfectly elastic nor perfectly viscid, and therefore an accurate formulation of the problem requires the study of an operator which interpolates between the "derivative of order zero" appearing in (1.46) and the "derivative of order one" appearing in (1.47), and of course fractional derivatives seem to perfectly fit such a scope.

From the point of view of the notation, in the description of fluids, the displacement function ε typically represents the “strain,” while the normalized force function σ typically represents the “stress” acting on the fluid particles.

To give a concrete feeling of this superposition of elastic and viscid effects, we recall here a purely mechanical model which was proposed in [67] (we actually simplify the discussion presented in [67], since we do not aim here at fully justified general statements).

The idea proposed by [67] is to take into account a system of springs, which react to forces elastically according to Hooke’s Law, and dashpots, or viscous dampers, in which strains and stresses are related by Newton’s Law.

To clarify the setting we will schematically draw springs and dashpots as described in Figure 1.5. Following [67], the model that we discuss here consists of a ladder-like structure with springs along one of the struts and dashpots on the rungs of the ladder; see Figure 1.6. The system contains n springs and $n-1$ dashpots and we will formally consider the limit as $n \to +\infty$. The superindices d and s will refer to the dashpots and the springs, respectively, while the subindices will refer to the position in the ladder. In particular, we can consider the elongation of the springs, denoted by $\varepsilon_0^s, \dots, \varepsilon_{n-1}^s$, and the ones of the dashpots, denoted by $\varepsilon_0^d, \dots, \varepsilon_{n-2}^d$. From Figure 1.6, we see that

$$\varepsilon_k^d = \varepsilon_{k+1}^s + \varepsilon_{k+1}^d. \tag{1.50}$$

Moreover, if $\sigma_1^s, \dots, \sigma_n^s$ denote the stresses on the springs and $\sigma_1^d, \dots, \sigma_{n-1}^d$ the ones on the dashpots, the parallel arrangement on the ladder gives

$$\sigma_k^s = \sigma_{k+1}^s + \sigma_k^d. \tag{1.51}$$

Also, the total elongation of the system is given by

$$\varepsilon = \varepsilon_0^s + \varepsilon_0^d, \tag{1.52}$$

and the stress at the end of the ladder is given by the one of the first spring, namely,

$$\sigma = \sigma_0^s. \tag{1.53}$$

We will consider springs with the same elastic coefficient, which we normalize in such a way that Hooke’s Law writes in this case as

$$\varepsilon_k^s = 2\sigma_k^s. \tag{1.54}$$

Also, the dashpots will be taken to be all with the same viscous coefficients and, in this way, Newton’s Law will be written as

$$\dot{\varepsilon}_k^d = \sigma_k^d. \tag{1.55}$$

Figure 1.5: Schematic representation of a spring (left) and a dashpot (right).

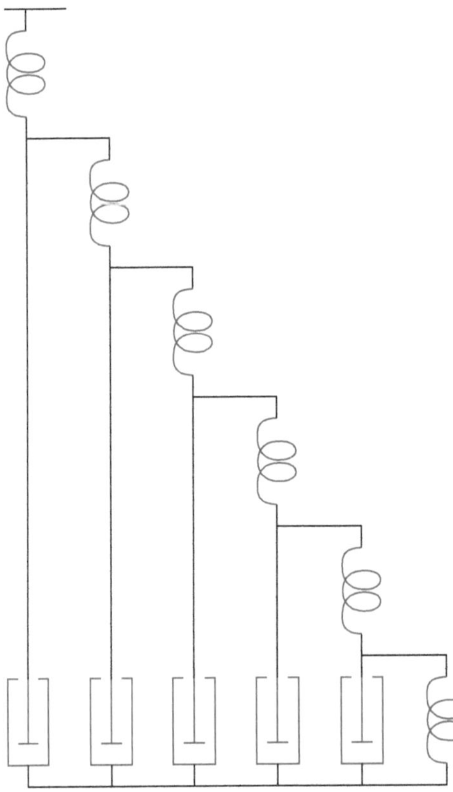

Figure 1.6: The spring–dashpot ladder of Example 1.7.

The case of different springs and dashpots can also be taken into account and it would produce a quantitatively different analysis (see [67] for details), but even this simpler case in which all the mechanical elements are the same produces some interesting effects that we now analyze.

First of all, by (1.50) and (1.54),

$$\varepsilon_k^d = 2\sigma_{k+1}^s + \varepsilon_{k+1}^d. \tag{1.56}$$

Similarly, by (1.51) and (1.55),

$$\sigma_k^s = \sigma_{k+1}^s + \dot{\varepsilon}_k^d. \tag{1.57}$$

It is now appropriate to consider the Laplace transform of a function u, which we denote by

$$\bar{u}(\omega) := \int_0^{+\infty} u(t)e^{-t\omega}\,dt.$$

For further reference, we observe that

$$\begin{aligned}\bar{\dot{u}}(\omega) &= \int_0^{+\infty} \dot{u}(t)e^{-t\omega}\,dt\\ &= \int_0^{+\infty} \left(\frac{d}{dt}(u(t)e^{-t\omega}) + \omega u(t)e^{-t\omega}\right) dt\\ &= -u(0) + \omega\bar{u}(\omega).\end{aligned} \tag{1.58}$$

In a similar way, considering the fractional derivative notation

$$D_{t,0}^{1/2}u(t) := \int_0^t \frac{\dot{u}(\tau)}{\sqrt{t-\tau}}\,d\tau,$$

to be compared with the general setting in the forthcoming formula (2.6), we have

$$\begin{aligned}\overline{D_{t,0}^{1/2}u}(\omega) &= \int_0^{+\infty}\left[\int_0^t \frac{\dot{u}(\tau)}{\sqrt{t-\tau}}\,d\tau\right]e^{-t\omega}\,dt\\ &= \int_0^{+\infty}\left[\dot{u}(\tau)\int_\tau^{+\infty}\frac{e^{-t\omega}}{\sqrt{t-\tau}}\,dt\right]d\tau\\ &= \int_0^{+\infty}\left[\dot{u}(\tau)e^{-\tau\omega}\int_0^{+\infty}\frac{e^{-\zeta\omega}}{\sqrt{\zeta}}\,d\zeta\right]d\tau\\ &= \frac{1}{\sqrt{\omega}}\int_0^{+\infty}\left[\dot{u}(\tau)e^{-\tau\omega}\int_0^{+\infty}\frac{e^{-\mu}}{\sqrt{\mu}}\,d\mu\right]d\tau\\ &= \frac{C}{\sqrt{\omega}}\int_0^{+\infty}\dot{u}(\tau)e^{-\tau\omega}\,d\tau\\ &= \frac{C}{\sqrt{\omega}}\int_0^{+\infty}\left(\frac{d}{d\tau}(u(\tau)e^{-\tau\omega}) + \omega u(\tau)e^{-\tau\omega}\right)d\tau\end{aligned}$$

$$= -\frac{Cu(0)}{\sqrt{\omega}} + C\sqrt{\omega} \int_0^{+\infty} u(\tau)e^{-\tau\omega}\,d\tau$$

$$= -\frac{Cu(0)}{\sqrt{\omega}} + C\sqrt{\omega}\bar{u}(\omega), \tag{1.59}$$

for a suitable $C > 0$.

Now, taking the Laplace transform of (1.56), we find

$$\bar{\varepsilon}_k^d = 2\bar{\sigma}_{k+1}^s + \bar{\varepsilon}_{k+1}^d,$$

and therefore

$$\frac{\bar{\varepsilon}_k^d}{2\bar{\sigma}_{k+1}^s} = 1 + \frac{\bar{\varepsilon}_{k+1}^d}{2\bar{\sigma}_{k+1}^s}. \tag{1.60}$$

Instead, taking the Laplace transform of (1.57) and recalling (1.58), assuming that the initial displacement vanishes, we find

$$\bar{\sigma}_k^s = \bar{\sigma}_{k+1}^s + \omega\bar{\varepsilon}_k^d. \tag{1.61}$$

We write this identity as

$$\bar{\sigma}_{k+1}^s = \bar{\sigma}_{k+2}^s + \omega\bar{\varepsilon}_{k+1}^d$$

and we substitute it in the right hand side of (1.60), concluding that

$$\begin{aligned}
\rho_k &:= \frac{\bar{\varepsilon}_k^d}{2\bar{\sigma}_{k+1}^s} \\
&= 1 + \frac{\bar{\varepsilon}_{k+1}^d}{2(\bar{\sigma}_{k+2}^s + \omega\bar{\varepsilon}_{k+1}^d)} \\
&= 1 + \frac{\rho_{k+1}}{1 + \omega\frac{\bar{\varepsilon}_{k+1}^d}{\bar{\sigma}_{k+2}^s}} \\
&= 1 + \frac{\rho_{k+1}}{1 + 2\omega\rho_{k+1}} \\
&= 1 + \frac{1}{2\omega + \frac{1}{\rho_{k+1}}}.
\end{aligned} \tag{1.62}$$

We can iterate (1.62) and then find

$$\rho_k = 1 + \cfrac{1}{2\omega + \cfrac{1}{1+\cfrac{1}{2\omega+\cfrac{1}{\rho_{k+2}}}}} = 1 + \cfrac{1}{2\omega + \cfrac{1}{1+\cfrac{1}{2\omega+\cfrac{1}{1+\cfrac{1}{2\omega+\cfrac{1}{\rho_{k+3}}}}}}},$$

and so on. That is, in the formal limit of infinitely many springs and dashpots,

$$\frac{\bar{\varepsilon}_0^d}{2\bar{\sigma}_1^s} = \rho_0 = 1 + \cfrac{1}{2\omega + \cfrac{1}{1+\cfrac{1}{2\omega+\cfrac{1}{\rho_2}}}} = 1 + \cfrac{1}{2\omega + \cfrac{1}{1+\cfrac{1}{2\omega+\cfrac{1}{1+\cfrac{1}{2\omega+\cfrac{1}{\ddots}}}}}}, \tag{1.63}$$

which is an infinite continuous fraction.

We observe that

$$\text{the right hand side of (1.63) is equal to } \frac{1+\sqrt{1+\frac{2}{\omega}}}{2}. \tag{1.64}$$

Indeed, the right hand side of (1.63) is a positive number, say X, and it satisfies

$$X = 1 + \frac{1}{2\omega + \frac{1}{X}}.$$

Solving X in this relation, we obtain (1.64), as desired.

Then, from (1.63) and (1.64), we conclude that

$$\frac{\bar{\varepsilon}_0^d}{2\bar{\sigma}_1^s} = \frac{1+\sqrt{1+\frac{2}{\omega}}}{2}. \tag{1.65}$$

On the other hand, recalling (1.52), (1.53), (1.54), and (1.61),

$$\begin{aligned}\frac{\bar{\varepsilon}(\omega)}{2\bar{\sigma}(\omega)} &= \frac{\bar{\varepsilon}_0^s + \bar{\varepsilon}_0^d}{2\bar{\sigma}_0^s} \\ &= \frac{\bar{\varepsilon}_0^s}{2\bar{\sigma}_0^s} + \frac{\bar{\varepsilon}_0^d}{2\bar{\sigma}_0^s} \\ &= 1 + \frac{\bar{\sigma}_1^s}{\bar{\sigma}_1^s + \omega\bar{\epsilon}_0^d} \times \frac{\bar{\varepsilon}_0^d}{2\bar{\sigma}_1^s},\end{aligned}$$

which combined with (1.65) gives

$$\frac{\bar{\varepsilon}(\omega)}{2\bar{\sigma}(\omega)} = 1 + \frac{\bar{\sigma}_1^s}{\bar{\sigma}_1^s + \omega\bar{\epsilon}_0^d} \times \frac{1+\sqrt{1+\frac{2}{\omega}}}{2}. \tag{1.66}$$

For small ω, we see that (1.66) becomes

$$\frac{\bar{\varepsilon}(\omega)}{\bar{\sigma}(\omega)} \simeq \sqrt{\frac{2}{\omega}}. \tag{1.67}$$

We observe that, at least formally, the regime of small ω corresponds to that of large t: a rigorous justification of these asymptotics is likely to rely on a suitable use of an

Abelian–Tauberian Theorem (see, e. g., [53]) and goes well beyond the scope of our heuristic argument (but see formulas (38) and (39) in [67] for a quantitative analysis of these asymptotics).

Hence, from (1.67), taking the initial displacement to be null, i. e., assuming that $\varepsilon(0) = 0$, and normalizing the constants to be unitary for the sake of simplicity, in light of (1.59), we can write, for small ω,

$$\bar{\sigma}(\omega) \simeq \sqrt{\omega}\bar{\varepsilon}(\omega) = \overline{D^{1/2}_{t,0}\varepsilon}(\omega),$$

and therefore, for large t, we have

$$D^{1/2}_{t,0}\varepsilon(t) \simeq \sigma(t).$$

This provides a heuristic, but rather convincing, motivation showing how fractional derivatives naturally surface in complex mechanical models involving springs and dashpots and suggesting that similar phenomena can arise in models in which both elastic and viscid effects contribute effectively to the macroscopic behavior of the system.

Example 1.8 (Diffusion along a comb structure). In this example we show how the geometry of the diffusion media can naturally produce fractional equations. We take into account a diffusion model along a comb structure. The diffusion along the backbone of the comb can be either classical or of space-fractional type (the model that we present was indeed introduced in [9] in the case of classical diffusion along the backbone, but we will present here an even more general setting that comprises space-fractional diffusion as well).

The ramified medium that we take into account in this example is a "comb," with a backbone on the horizontal axis and a fine grid of vertical fingers which are located at mutual distance $\epsilon > 0$; see Figure 1.7.

We consider the superposition of (i) a horizontal diffusion along the backbone driven by $-(-\Delta)^s_x$, for some $s \in (0,1]$, the case $s = 1$ corresponding to classical diffusion and the case $s \in (0,1)$ to space-fractional diffusion as in (1.18) and (1.19) and (ii) a vertical diffusion along the fingers of classical type. To make the model attain a significant limit in the continuous approximation in which $\epsilon \searrow 0$, since the fingers are infinitely many and their distance tends to zero, it is convenient to assume that the vertical diffusion is subject to a diffusive coefficient of order ϵ. The initial position is taken for simplicity to be a concentrated mass at the origin. This model translates into the following mathematical formulation:

$$\begin{cases} \partial_t u(x,y,t) = -\delta_0(y)(-\Delta)^s_x u(x,y,t) + \epsilon \sum_{j\in\mathbb{Z}} \delta_0(\epsilon j)\partial^2_y u(x,y,t) \\ \quad \text{for all } (x,y) \in \mathbb{R}^2 \text{ and } t > 0, \\ u(x,y,0) = \delta_0(x)\delta_0(y). \end{cases} \tag{1.68}$$

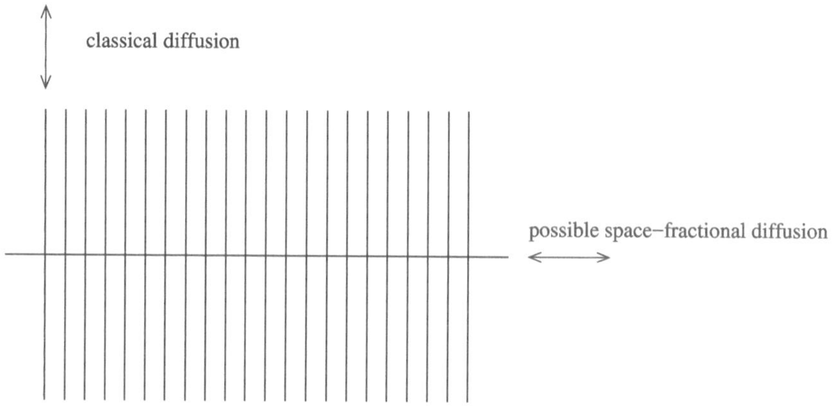

Figure 1.7: The comb structure along which diffusion takes place in Example 1.8.

As customary, the notation δ_0 denotes here the Dirac Delta function centered at the origin. We point out that

$$\lim_{\epsilon\searrow 0} \epsilon \sum_{j\in\mathbb{Z}} \delta_0(\epsilon j) = 1, \tag{1.69}$$

in the distributional sense. Indeed, if $\varphi \in C_0^\infty(\mathbb{R})$,

$$\epsilon \sum_{j\in\mathbb{Z}} \int_{\mathbb{R}} \delta_0(\epsilon j)\varphi(y)\,dy = \epsilon \sum_{j\in\mathbb{Z}} \varphi(\epsilon j).$$

Since the latter can be seen as a Riemann sum, we find

$$\lim_{\epsilon\searrow 0} \epsilon \sum_{j\in\mathbb{Z}} \int_{\mathbb{R}} \delta_0(\epsilon j)\varphi(y)\,dy = \int_{\mathbb{R}} \varphi(y)\,dy,$$

which gives (1.69).

Hence, by (1.69), one can take into account the continuous limit of (1.68) as $\epsilon \searrow 0$, which we can write in the form

$$\begin{cases} \partial_t u(x,y,t) = -\delta_0(y)(-\Delta)_x^s u(x,y,t) + \partial_y^2 u(x,y,t) & \text{for all } (x,y) \in \mathbb{R}^2 \text{ and } t > 0, \\ u(x,y,0) = \delta_0(x)\delta_0(y). \end{cases} \tag{1.70}$$

We consider the effective transport along the backbone $\{y = 0\}$, given by the function

$$U(x,t) := \int_{\mathbb{R}} u(x,y,t)\,dy. \tag{1.71}$$

Our claim is that U satisfies a time-fractional equation given by

$$\begin{cases} D_{t,0}^{1/2} U(x,t) = -(-\Delta)_x^s U(x,t) & \text{for all } x \in \mathbb{R} \text{ and } t > 0, \\ U(x,0) = \delta_0(x). \end{cases} \tag{1.72}$$

Here, up to normalization constants, we take

$$D^{1/2}_{t,0}U(x,t) := \int_0^t \frac{\partial_t U(x,\tau)}{\sqrt{t-\tau}}\,d\tau$$
$$= \int_0^t \frac{\partial_t U(x,t-\tau)}{\sqrt{\tau}}\,d\tau,$$

and one can compare this expression with the similar ones in (1.3) and (1.5), as well as with the more general setting that we will introduce in (2.6).

To prove (1.72), we first check the initial condition. For this, using (1.70), we see that

$$U(x,0) = \int_{\mathbb{R}} u(x,y,0)\,dy = \int_{\mathbb{R}} \delta_0(x)\delta_0(y)\,dy = \delta_0(x). \tag{1.73}$$

Having checked the initial condition in (1.72), we now aim at proving the validity of the evolution equation in (1.72). To this end, it is convenient to consider the Fourier–Laplace transform of a function $v = v(x,t)$, namely, the Fourier transform in the variable x combined with the Laplace transform in the variable t. Namely, up to normalization constants that we omit, we define

$$\mathcal{E}_v(\xi,\omega) := \iint_{\mathbb{R}\times(0,+\infty)} v(x,t)e^{-ix\xi-t\omega}\,dx\,dt.$$

We observe that, for every $x \in \mathbb{R}$,

$$\int_0^{+\infty} D^{1/2}_{t,0}U(x,t)e^{-t\omega}\,dt$$
$$= \int_0^{+\infty}\left[\int_0^t \frac{\partial_t U(x,t-\tau)e^{-t\omega}}{\sqrt{\tau}}\,d\tau\right]dt$$
$$= \int_0^{+\infty}\left[\int_\tau^{+\infty} \frac{\partial_t U(x,t-\tau)e^{-t\omega}}{\sqrt{\tau}}\,dt\right]d\tau$$
$$= \int_0^{+\infty}\left[\int_\tau^{+\infty} (\partial_t(U(x,t-\tau)e^{-t\omega}) + \omega U(x,t-\tau)e^{-t\omega})\,dt\right]\frac{d\tau}{\sqrt{\tau}}$$
$$= \int_0^{+\infty}\left[-U(x,0)e^{-\tau\omega} + \int_\tau^{+\infty} \omega U(x,t-\tau)e^{-t\omega}\,dt\right]\frac{d\tau}{\sqrt{\tau}}$$
$$= -\sqrt{\frac{\pi}{\omega}}U(x,0) + \omega\int_0^{+\infty}\left[\int_0^{+\infty} U(x,\sigma)e^{-(\sigma+\tau)\omega}\,d\sigma\right]\frac{d\tau}{\sqrt{\tau}}$$
$$= -\sqrt{\frac{\pi}{\omega}}U(x,0) + \sqrt{\pi\omega}\int_0^{+\infty} U(x,\sigma)e^{-\sigma\omega}\,d\sigma.$$

As a consequence,

$$\begin{aligned}\mathcal{E}_{D^{1/2}_{t,0}U}(\xi,\omega) &= \int_{\mathbb{R}} \left[-\sqrt{\frac{\pi}{\omega}} U(x,0) + \sqrt{\pi\omega} \int_0^{+\infty} U(x,\sigma) e^{-\sigma\omega}\, d\sigma \right] e^{-ix\xi}\, dx \\ &= -\sqrt{\frac{\pi}{\omega}} \hat{U}(\xi,0) + \sqrt{\pi\omega}\,\mathcal{E}_U(\xi,\omega), \end{aligned} \tag{1.74}$$

where $\hat{U}$ denotes the Fourier transform of U in the variable x.

Moreover, by (1.18), up to normalizing constants we can write

$$\begin{aligned}\mathcal{E}_{(-\Delta)^s_x U}(\xi,\omega) &= \int_0^{+\infty} |\xi|^{2s} \hat{U}(\xi,t) e^{-t\omega}\, dt \\ &= |\xi|^{2s} \mathcal{E}_U(\xi,\omega).\end{aligned}$$

By this and (1.74), we see that, to check the validity of the evolution equation in (1.72), recalling (1.73), up to normalizing constants we need to establish that

$$|\xi|^{2s}\mathcal{E}_U(\xi,\omega) - \sqrt{\frac{1}{\omega}} + \sqrt{\omega}\,\mathcal{E}_U(\xi,\omega) = 0. \tag{1.75}$$

To this end, we consider the Fourier–Laplace transform of the function u that solves (1.70). Namely, we define

$$W(\xi,y,\omega) := \mathcal{E}_u(\xi,y,\omega) = \iint_{\mathbb{R}\times(0,+\infty)} u(x,y,t) e^{-ix\xi - t\omega}\, dx\, dt. \tag{1.76}$$

In view of (1.70), we point out that

$$\begin{aligned}&-\delta_0(y)|\xi|^{2s} W(\xi,y,\omega) + \partial^2_y W(\xi,y,\omega) \\ &\quad= -\delta_0(y) \iint_{\mathbb{R}\times(0,+\infty)} |\xi|^{2s} u(x,y,t) e^{-ix\xi - t\omega}\, dx\, dt + \partial^2_y W(\xi,y,\omega) \\ &\quad= -\delta_0(y) \int_0^{+\infty} |\xi|^{2s} \hat{u}(\xi,y,t) e^{-t\omega}\, dt + \partial^2_y W(\xi,y,\omega) \\ &\quad= \mathcal{E}_{-\delta_0(y)(-\Delta)^s_x u + \partial^2_y u}(\xi,y,\omega) \\ &\quad= \mathcal{E}_{\partial_t u}(\xi,y,\omega) \\ &\quad= \iint_{\mathbb{R}\times(0,+\infty)} \partial_t u(x,y,t) e^{-ix\xi - t\omega}\, dx\, dt \\ &\quad= \iint_{\mathbb{R}\times(0,+\infty)} \left(\partial_t\left(u(x,y,t)e^{-ix\xi-t\omega}\right) + \omega u(x,y,t) e^{-ix\xi - t\omega}\right) dx\, dt\end{aligned}$$

$$= -\int_{\mathbb{R}} u(x,y,0)e^{-ix\xi}\,dx + \omega \iint_{\mathbb{R}\times(0,+\infty)} u(x,y,t)e^{-ix\xi-t\omega}\,dx\,dt$$
$$= -\delta_0(y) + \omega W(\xi,y,\omega).$$

Hence the map $\mathbb{R} \ni y \mapsto W(\xi,y,\omega)$ satisfies

$$\partial_y^2 W = a^2 W - b(2a+c)\delta_0(y) + c\delta_0(y)W, \tag{1.77}$$

where $a = a(\omega) := \sqrt{\omega}$, $b = b(\xi,\omega) := \frac{1}{2\sqrt{\omega}+|\xi|^{2s}}$ and $c = c(\xi) := |\xi|^{2s}$.

Hence, fixing $\xi \in \mathbb{R}$ and $t > 0$, one considers (1.77) as an equation for a function of $y \in \mathbb{R}$, in which a, b, and c are coefficients independent of y.

With this in mind, we observe that if $a > 0$, $b \in \mathbb{R}$, and $g(y) := be^{-a|y|}$, we have

$$g''(y) = a^2 g(y) - b(2a+c)\delta_0(y) + c\delta_0(y)g(y), \tag{1.78}$$

for any $c \in \mathbb{R}$. To check this, let $\varphi \in C_0^\infty(\mathbb{R})$. Then, we compute

$$\int_{\mathbb{R}} (g''(y) - a^2 g(y) + b(2a+c)\delta_0(y) - c\delta_0(y)g(y))\varphi(y)\,dy$$
$$= \int_{\mathbb{R}} g(y)\varphi''(y)\,dy - a^2 \int_{\mathbb{R}} g(y)\varphi(y)\,dy + b(2a+c)\varphi(0) - cg(0)\varphi(0)$$
$$= b\int_{\mathbb{R}} e^{-a|y|}\varphi''(y)\,dy - a^2 b\int_{\mathbb{R}} e^{-a|y|}\varphi(y)\,dy + b(2a+c)\varphi(0) - bc\varphi(0)$$
$$= b\int_0^{+\infty} e^{-ay}\varphi''(y)\,dy + b\int_{-\infty}^0 e^{ay}\varphi''(y)\,dy$$
$$-a^2 b\int_0^{+\infty} e^{-ay}\varphi(y)\,dy - a^2 b\int_{-\infty}^0 e^{ay}\varphi(y)\,dy$$
$$+b(2a+c)\varphi(0) - bc\varphi(0)$$
$$= -b\varphi'(0) + ab\int_0^{+\infty} e^{-ay}\varphi'(y)\,dy + b\varphi'(0) - ab\int_{-\infty}^0 e^{ay}\varphi'(y)\,dy$$
$$-a^2 b\int_0^{+\infty} e^{-ay}\varphi(y)\,dy - a^2 b\int_{-\infty}^0 e^{ay}\varphi(y)\,dy$$
$$+b(2a+c)\varphi(0) - bc\varphi(0)$$
$$= -b\varphi'(0) - ab\varphi(0) + a^2 b\int_0^{+\infty} e^{-ay}\varphi(y)\,dy$$
$$+b\varphi'(0) - ab\varphi(0) + a^2 b\int_{-\infty}^0 e^{ay}\varphi(y)\,dy$$

$$
\begin{aligned}
&-a^2b\int_0^{+\infty} e^{-ay}\varphi(y)\,dy - a^2b\int_{-\infty}^0 e^{ay}\varphi(y)\,dy\\
&+b(2a+c)\varphi(0) - bc\varphi(0)\\
&= 0,
\end{aligned}
$$

which proves (1.78).

In light of (1.78), we can write a solution of (1.77) in the form

$$
W(\xi, y, \omega) = b(\xi, \omega)e^{-a(\omega)|y|} = \frac{e^{-\sqrt{\omega}|y|}}{2\sqrt{\omega} + |\xi|^{2s}}.
$$

As a consequence, recalling (1.71) and (1.76),

$$
\begin{aligned}
\frac{2}{\sqrt{\omega}(2\sqrt{\omega} + |\xi|^{2s})} &= \int_{\mathbb{R}} \frac{e^{-\sqrt{\omega}|y|}}{2\sqrt{\omega} + |\xi|^{2s}}\,dy\\
&= \int_{\mathbb{R}} W(\xi, y, \omega)\,dy\\
&= \mathcal{E}_U(\xi, \omega).
\end{aligned}
$$

Hence, we see that

$$
(|\xi|^{2s} + 2\sqrt{\omega})\mathcal{E}_U(\xi, \omega) = \frac{2}{\sqrt{\omega}}.
$$

With this, we have checked the validity of (1.75) and thus we have established the fractional equation in (1.72).

Example 1.9 (Fractional operators and option pricing). One of the most common features in mathematical finance is option pricing. Suppose that a holder wants to buy some good, say, the ticket to the final match of the Champions League (an "underlying" in financial jargon), which will occur at time T.

The price of this ticket depends on time, and say that at time $t \in [0, T]$ this ticket costs S_t (we are using here a common notation in finance by using the subindex t to denote the value of a function at a given time t). Rather than buying directly the ticket for the price S_t, he/she can buy an option[2] that allow him/her to buy the ticket at time T for a fixed price K (called in jargon "strike price").

The question is: should the holder buy the ticket or the option? Or, more precisely: what is the value that such an option has in the market?

2 For simplicity, we are discussing here the case of *European options*, in which the holder can only employ the option exactly at a given time T. Instead, in *American options* the holder has the right, but not the obligation, to buy the underlying at an agreed-upon price at any time $t \in [0, T]$ (and, of course, the seller has an obligation to sell the underlying if the option is exercised).

We suppose that the value V of the option depends on the time t and on the cost of the ticket S, so we will write $V = V(S,t)$. What is obvious is the final value of the option, since

$$V(S,T) := \max\{S-K,0\}. \tag{1.79}$$

Indeed, at the final time, one can decide to either buy the ticket at a certain price S or the option at price $V(S,T)$ and then the ticket at the strike price K, and these two operations should be equivalent, hence $S = V(S,T)+K$ (this if $S \geq K$, otherwise the option has simply no value, since anyone can just buy the ticket for a more convenient price, hence confirming (1.79)).

Typically, in order to determine the option price, one can prove that it solves some partial differential equation. For instance, one can follow the Black–Scholes model and a variation of it; see, e. g., [13]. In our framework, a risk-neutral dynamic of the asset (the price of the ticket) is given by an exponential model

$$S_t = S_0 \exp(\mu t + \sigma W_t),$$

where $\mu \in \mathbb{R}$ denotes a drift which measures the expected yield of the underlying, $\sigma \in (0,+\infty)$ is a diffusion coefficient which measures the degree of variation of the trading price of the asset (in jargon, "volatility"), and W_t is a "reasonable" stochastic process modeling the unpredictable oscillations of the market. In the classical case, W_t was taken to be simply Brownian motion, but recently fractional Brownian motions and jump processes have been taken into account to model possibly different evolutions of the market.

Without going into technical details, following, e. g., [73], we will suppose that the stochastic motion arises from the superposition of a Brownian motion with a jump process. Concretely, we assume that the "infinitesimal generator" of the process is of the form

$$\mathcal{A} := a\partial^2 - b(-\Delta)^s, \quad \text{for some } s \in (0,1), \tag{1.80}$$

with a, $b \geq 0$, that is, if we denote by $\mathbb{E}$ the expected value, we assume that

$$\lim_{\tau\to 0} \frac{\mathbb{E}(f(S_{t+\tau})) - f(S_t)}{\tau} = \mathcal{A}f(S_t). \tag{1.81}$$

The heuristic rationale of this formula (say, with $t = 0$) is as follows: in the appropriate scale, the probability density of a particle traveling under the superposition of a Brownian motion with a jump process satisfies a heat equation in which the classical Laplacian is replaced by the operator $\mathcal{A}$ (see, e. g., [73]), and hence, if $S_{t,x}$ denotes the stochastic evolution starting at x (i. e., $S_{0,x} = x$) and $u(x,t) := \mathbb{E}(f(S_{t,x}))$, we expect that u is a solution of

$$\begin{cases} \partial_t u(x,t) = \mathcal{A}u(x,t) & \text{if } t > 0, \\ u(x,0) = f(x). \end{cases}$$

Hence we can expect that

$$\begin{aligned}\lim_{\tau\to 0}\frac{\mathbb{E}(f(S_{\tau,x}))-f(S_{0,x})}{\tau} &= \lim_{\tau\to 0}\frac{u(x,\tau)-f(x)}{\tau}\\ &= \lim_{\tau\to 0}\frac{u(x,\tau)-u(x,0)}{\tau} = \partial_t u(x,0)\\ &= \mathcal{A}u(x,0) = \mathcal{A}f(x)\\ &= \mathcal{A}f(S_{0,x}),\end{aligned}$$

which can provide a justification for (1.81).

We also introduce a suitable Itô formula, of the form

$$\frac{d}{dt}f(S_t) = f'(S_t)\,\frac{dS_t}{dt} + \mathcal{A}f(S_t). \tag{1.82}$$

To try to justify (1.82), one can note that the stochastic process can move "indifferently" up or down, say, $S_{t+\tau} - S_t$ has no "definite sign," hence

$$\lim_{\tau\to 0}\frac{\mathbb{E}(f'(S_t)(S_{t+\tau}-S_t))}{\tau} = 0.$$

That is, by formal Taylor expansion and recalling (1.81),

$$\begin{aligned}\mathcal{A}f(S_t) &= \lim_{\tau\to 0}\frac{\mathbb{E}(f(S_{t+\tau}))-f(S_{t+\tau})}{\tau} + \frac{f(S_{t+\tau})-f(S_t)}{\tau}\\ &= \lim_{\tau\to 0}\frac{\mathbb{E}(f(S_t)+f'(S_t)(S_{t+\tau}-S_t)) - (f(S_t)+f'(S_t)(S_{t+\tau}-S_t))}{\tau} + \frac{d}{dt}f(S_t)\\ &= \lim_{\tau\to 0} -\frac{f'(S_t)(S_{t+\tau}-S_t)}{\tau} + \frac{d}{dt}f(S_t)\\ &= -f'(S_t)\,\frac{dS_t}{dt} + \frac{d}{dt}f(S_t),\end{aligned}$$

which gives a heuristic justification of (1.82).

Clearly, formula (1.82) can be extended to functions which also depend explicitly on time, thus yielding

$$\frac{d}{dt}f(S_t,t) = \frac{\partial f}{\partial t}(S_t,t) + \frac{\partial f}{\partial S}(S_t,t)\,\frac{dS_t}{dt} + \mathcal{A}f(S_t,t). \tag{1.83}$$

Now, the idea is to try to determine an equation for the value V of the option, under reasonable assumptions on the market. It would be desirable to release V from any randomness and uncertainty offered by the market. In a sense, the oscillations of the market could affect the price S_t of the ticket at time t, but we would like to know $V = V(S,t)$ in a way which is independent of this randomness (only in dependence of the time t and any possible price of the ticket S), and then evaluate $V(S_t,t)$ to have the value of the option at time t, for the price S_t of the ticket.

To do so, we try to build a "risk-free" portfolio. Given the oscillations of the market, the strategy of possessing only a certain number of options is highly subject to the market uncertainties and it would be safer for a holder, or for a company, not only to buy or sell one or more options but also to buy or sell one or more tickets for the Champions League final. That is, a suitable portfolio should be of the form

$$P(S,t) = V(S,t) + \delta S, \tag{1.84}$$

with $\delta \in \mathbb{R}$ giving the number of tickets to possess with respect to the option to make the total portfolio as "stable" as possible (negative values of δ would correspond to selling tickets, rather than buying them). To choose δ in such a way that P becomes as close to risk-free as possible, it is desirable to reduce the oscillations of P with respect to the variations of S. To this end, we make the simplifying assumption that $V(0,t) = 0$, i. e., the value of the option is null if the price of the ticket is null as well, and we observe that

$$V(S,t) = V(S,t) - V(0,t) = \frac{\partial V}{\partial S}(0,t)S + O(S^2),$$

that is,

$$V(S,t) - \frac{\partial V}{\partial S}(0,t)S = O(S^2).$$

Comparing this with (1.84), we see that a reasonable possibility to make the portfolio as independent as possible of the fluctuating value S is to choose $\delta := -\frac{\partial V}{\partial S}(0,t)$ in (1.84), which leads to

$$P(S,t) = V(S,t) - \frac{\partial V}{\partial S}(0,t)S. \tag{1.85}$$

Let us make a brief comment on the minus sign in (1.85). One can suppose, for instance, that V is monotone increasing with respect to S (the more the ticket costs, the more valuable is the option) and thus $\frac{\partial V}{\partial S}(0,t) > 0$. In the setting of (1.85), this means that, to maintain a balanced portfolio, if one buys options it is appropriate to sell tickets.

One then makes the ansatz that (1.85) gives indeed a risk-free portfolio. Under this assumption, the time evolution of P is just governed by the interest rate and one can write that the time evolution of $P(S_t,t)$ is simply $P_0 e^{rt}$, for some r, $P_0 \in \mathbb{R}$. As a consequence,

$$\frac{d}{dt}P(S_t,t) = \frac{d}{dt}(P_0 e^{rt}) = rP_0 e^{rt} = rP(S_t,t).$$

Hence, recalling (1.85),

$$\frac{d}{dt}\left(V(S_t,t) - \frac{\partial V}{\partial S}(0,t)S_t\right) = rV(S_t,t) - r\frac{\partial V}{\partial S}(0,t)S_t. \tag{1.86}$$

On the other hand, by (1.83),

$$\frac{d}{dt}V(S_t,t) = \frac{\partial V}{\partial t}(S_t,t) + \frac{\partial V}{\partial S}(S_t,t)\,\frac{dS_t}{dt} + \mathcal{A}V(S_t,t),$$

and so (1.86) becomes

$$\begin{aligned}&\frac{\partial V}{\partial t}(S_t,t) + \frac{\partial V}{\partial S}(S_t,t)\,\frac{dS_t}{dt} + \mathcal{A}V(S_t,t) - \frac{d}{dt}\Big(\frac{\partial V}{\partial S}(0,t)S_t\Big)\\ &= rV(S_t,t) - r\frac{\partial V}{\partial S}(0,t)S_t.\end{aligned} \tag{1.87}$$

It is also customary to neglect the dependence of $\frac{\partial V}{\partial S}(0,t)$ on t (say, the values of the options of very cheap tickets are more or less the same independent of time), and thus replace the term $\frac{d}{dt}(\frac{\partial V}{\partial S}(0,t)S_t)$ with $\frac{\partial V}{\partial S}(0,t)\,\frac{dS_t}{dt}$. With this approximation, one obtains from (1.87), after simplifying two terms,

$$\frac{\partial V}{\partial t}(S_t,t) + \mathcal{A}V(S_t,t) = rV(S_t,t) - r\frac{\partial V}{\partial S}(0,t)S_t. \tag{1.88}$$

Recalling (1.80), when $b = 0$ one obtains from (1.88) the classical Black–Scholes equation

$$\frac{\partial V}{\partial t} + rS_t\frac{\partial V}{\partial S_t} + \frac{\sigma^2}{2}S_t^2\frac{\partial^2 V}{\partial S_t^2} - rV = 0 \quad \text{in } \mathbb{R}\times(0,T].$$

In general, when $b \neq 0$ (and possibly $a = 0$) one obtains in (1.88) a nonlocal evolution equation of fractional type. Such an equation is complemented with the terminal condition in (1.79).

Example 1.10 (Complex analysis and Hilbert transform). Given a (nice) function $u : \mathbb{R} \to \mathbb{R}$, the Hilbert transform of u is defined by

$$H_u(t) := -\frac{1}{\pi}\int_{-\infty}^{\infty} \frac{u(t)-u(\tau)}{t-\tau}\,d\tau.$$

We observe that

$$\int_{|t-\tau|\in(r,R)} \frac{u(t)}{t-\tau}\,d\tau = u(t)\int_{|\vartheta|\in(r,R)} \frac{d\vartheta}{\vartheta} = u(t)\Big(\int_r^R \frac{d\vartheta}{\vartheta} + \int_{-R}^{-r}\frac{d\vartheta}{\vartheta}\Big) = 0,$$

for all $0 < r < R$. Hence, in the principal value sense, after cancellations, one can also write

$$H_u(t) := \frac{1}{\pi}\int_{-\infty}^{\infty} \frac{u(\tau)}{t-\tau}\,d\tau.$$

Among the others, a natural application of the Hilbert transform occurs in complex analysis. Indeed, identifying points $(x,y) \in \mathbb{R} \times [0,+\infty)$ of the real upper half-plane with points $x+iy \in \{z \in \mathbb{C} \text{ with } \Im z \geq 0\}$ of the complex upper half-plane, *on the boundary of the half-plane, the two harmonic conjugate functions of a holomorphic function are related via the Hilbert transform*. More precisely, if f is holomorphic in the complex upper half-plane, we write $z = x + iy$ with $x \in \mathbb{R}$ and $y \geq 0$, and

$$f(z) = u(x,y) + iv(x,y).$$

We also set $u_0(x) := u(x,0)$ and $v_0(x) := v(x,0)$. Then, under natural regularity assumptions, we have

$$v_0(x) = H_{u_0}(x), \quad \text{for all } x \in \mathbb{R}. \tag{1.89}$$

For rigorous complex analysis results, we refer, e. g., to Theorem 93 on page 125 of [72] and to Theorems 3 and 4 on pages 77–78 of [54]. See also Sections 2.6–2.9 in [54] for a number of concrete applications of the Hilbert transform. We sketch two arguments to establish (1.89), one based on classical complex methods and one exploiting fractional calculus.

The first argument goes as follows. For any $z = x + iy$ with $y > 0$, we consider the Hilbert transform of u_0, up to normalization constants, defined with a complex variable, namely, we set

$$F(z) := \frac{1}{\pi i} \int_{\mathbb{R}} \frac{u_0(t)}{t-z}\, dt.$$

Then, F is holomorphic in the upper half-plane. Moreover,

$$\begin{aligned} F(z) &= \frac{1}{\pi i} \int_{\mathbb{R}} \frac{u_0(t)}{t-x-iy}\, dt \\ &= \frac{1}{\pi} \int_{\mathbb{R}} \frac{u_0(t)(i(x-t)+y)}{(t-x)^2+y^2}\, dt. \end{aligned} \tag{1.90}$$

We let $\tilde{u}(x,y) := \Re F(x+iy)$ and $\tilde{v}(x,y) := \Im F(x+iy)$. Then, using the substitution $w := (t-x)/y$,

$$\begin{aligned} \lim_{y \searrow 0} \tilde{u}(x,y) &= \lim_{y \searrow 0} \frac{1}{\pi} \int_{\mathbb{R}} \frac{u_0(t)y}{(t-x)^2+y^2}\, dt \\ &= \lim_{y \searrow 0} \frac{1}{\pi} \int_{\mathbb{R}} \frac{u_0(x+wy)}{w^2+1}\, dw \\ &= \frac{u_0(x)}{\pi} \int_{\mathbb{R}} \frac{dw}{w^2+1} \\ &= u_0(x). \end{aligned}$$

That is, the real part of F coincides with the harmonic extension of u_0 to the upper half-plane, up to harmonic functions vanishing on the trace. Therefore (reducing to finite energy solutions), we suppose that $\tilde{u} = u$. Since $\tilde{v}$ and v are the conjugate harmonic functions of $\tilde{u}$ and u, respectively, from the Cauchy–Riemann equations we thereby find

$$0 = \partial_y(\tilde{u} - u) = -\partial_x(\tilde{v} - v)$$
$$\text{and} \quad 0 = \partial_x(\tilde{u} - u) = \partial_y(\tilde{v} - v).$$

Hence (restricting to functions with finite energy), we have $\tilde{v} = v$.

From these observations and (1.90), we find

$$\begin{aligned} v_0(x) &= \lim_{y \searrow 0} \tilde{v}(x, y) \\ &= \lim_{y \searrow 0} \frac{1}{\pi} \int_{\mathbb{R}} \frac{u_0(t)(x - t)}{(t - x)^2 + y^2}\, dt \\ &= \lim_{y \searrow 0} \frac{1}{\pi} \int_{\mathbb{R}} \frac{(u_0(t) - u_0(x))(x - t)}{(t - x)^2 + y^2}\, dt \\ &= \frac{1}{\pi} \int_{\mathbb{R}} \frac{(u_0(t) - u_0(x))(x - t)}{(t - x)^2}\, dt \\ &= -\frac{1}{\pi} \int_{\mathbb{R}} \frac{u_0(t) - u_0(x)}{t - x}\, dt, \end{aligned}$$

which gives (1.89).

We now present another argument to establish (1.89) based on fractional calculus. Since u is harmonic in the upper half-plane, using the Fourier transform in the x variable, and dropping normalization constants for the sake of simplicity, we see that, for all $y \in (0, +\infty)$,

$$0 = \mathcal{F}(\Delta u)(\xi, y) = -|\xi|^2 \hat{u}(\xi, y) + \partial_y^2 \hat{u}(\xi, y),$$

and thus $\hat{u}(\xi, y) = C(\xi)e^{-|\xi|y}$, with $C(\xi) \in \mathbb{R}$. This gives

$$\partial_y \hat{u}(\xi, 0) = -|\xi| C(\xi) = -|\xi| \hat{u}(\xi, 0) = -|\xi| \hat{u}_0(\xi).$$

This and (1.18) give

$$\partial_y u(x, 0) = \mathcal{F}^{-1}(\partial_y \hat{u}(\xi, 0)) = -\mathcal{F}^{-1}(|\xi| \hat{u}_0(\xi)) = -\sqrt{-\Delta} u_0(x).$$

Hence, recalling (1.19), dropping normalization constants for the sake of simplicity, and using a parity cancellation, we see that

$$\begin{aligned}
\partial_y u(x,0) &= \frac{1}{2}\int_{\mathbb{R}} \frac{u_0(x+\theta)+u_0(x-\theta)-2u_0(x)}{\theta^2}\,d\theta \\
&= \lim_{\delta\searrow 0}\int_{\mathbb{R}\setminus(-\delta,\delta)} \frac{u_0(x+\theta)-u_0(x)}{\theta^2}\,d\theta \\
&= \lim_{\delta\searrow 0}\int_{\mathbb{R}\setminus(-\delta,\delta)} \frac{u_0(x+\theta)-u_0(x)-u_0'(x)\theta\Psi(\theta)}{\theta^2}\,d\theta \\
&= \int_{\mathbb{R}} \frac{u_0(x+\theta)-u_0(x)-u_0'(x)\theta\Psi(\theta)}{\theta^2}\,d\theta \\
&= \int_{\mathbb{R}} \frac{u_0(\tau)-u_0(x)-u_0'(x)(\tau-x)\Psi(\tau-x)}{(\tau-x)^2}\,d\tau,
\end{aligned}$$

with $\Psi \in C_0^\infty([-2,2],[0,1])$ being an even function such that $\Psi = 1$ in $[-1,1]$. Hence, from the Cauchy–Riemann equations,

$$-\partial_x v_0(x) = -\partial_x v(x,0) = \int_{\mathbb{R}} \frac{u_0(\tau)-u_0(x)-u_0'(x)(\tau-x)\Psi(\tau-x)}{(\tau-x)^2}\,d\tau.$$

As a consequence,

$$\begin{aligned}
-v_0(x) &= -\int_{-\infty}^{x} \partial_x v_0(t)\,dt \\
&= \int_{-\infty}^{x}\left[\int_{\mathbb{R}} \frac{u_0(\tau)-u_0(t)-u_0'(t)(\tau-t)\Psi(\tau-t)}{(\tau-t)^2}\,d\tau\right]dt \\
&= \lim_{R\to+\infty}\int_{-\infty}^{x}\left[\int_{-R+t}^{R+t} \frac{u_0(\tau)-u_0(t)-u_0'(t)(\tau-t)\Psi(\tau-t)}{(\tau-t)^2}\,d\tau\right]dt.
\end{aligned} \tag{1.91}$$

By an integration by parts, we note that

$$\begin{aligned}
&\int_{-\infty}^{x} \frac{u_0(\tau)-u_0(t)-u_0'(t)(\tau-t)\Psi(\tau-t)}{(\tau-t)^2}\,dt \\
&= \int_{-\infty}^{x} \big(u_0(\tau)-u_0(t)-u_0'(t)(\tau-t)\Psi(\tau-t)\big)\,\frac{d}{dt}\left(\frac{1}{\tau-t}\right)dt \\
&= \frac{u_0(\tau)-u_0(x)-u_0'(x)(\tau-x)\Psi(\tau-x)}{\tau-x}
\end{aligned}$$

$$+\int_{-\infty}^{x} \frac{u_0'(t) + u_0''(t)(\tau - t)\Psi(\tau - t) - u_0'(t)\Psi(\tau - t) - u_0'(t)(\tau - t)\Psi'(\tau - t)}{\tau - t}\,dt$$

$$= \frac{u_0(\tau) - u_0(x)}{\tau - x} - u_0'(x)\Psi(\tau - x)$$

$$+\int_{-\infty}^{x} \left(\frac{u_0'(t)}{\tau - t} + u_0''(t)\Psi(\tau - t) - \frac{u_0'(t)\Psi(\tau - t)}{\tau - t} - u_0'(t)\Psi'(\tau - t) \right) dt$$

$$= \frac{u_0(\tau) - u_0(x)}{\tau - x} - u_0'(x)\Psi(\tau - x)$$

$$+\int_{-\infty}^{x} \left(\frac{u_0'(t) - u_0'(t)\Psi(\tau - t)}{\tau - t} + \frac{d}{dt}\big(u_0'(t)\Psi(\tau - t)\big) \right) dt$$

$$= \frac{u_0(\tau) - u_0(x)}{\tau - x} + \int_{-\infty}^{x} u_0'(t)\Phi(\tau - t)\,dt,$$

where we defined the odd function

$$\Phi(r) := \frac{1 - \Psi(r)}{r}.$$

By inserting this information into (1.91) and exchanging integrals, one obtains

$$-v_0(x) = \lim_{R \to +\infty} \int_{-R+t}^{R+t} \left[\frac{u_0(\tau) - u_0(x)}{\tau - x} + \int_{-\infty}^{x} u_0'(t)\Phi(\tau - t)\,dt \right] d\tau$$

$$= \int_{\mathbb{R}} \frac{u_0(\tau) - u_0(x)}{\tau - x}\,d\tau + \lim_{R \to +\infty} \int_{-\infty}^{x} u_0'(t) \left[\int_{-R}^{R} \Phi(\theta)\,d\theta \right] dt$$

$$= \int_{\mathbb{R}} \frac{u_0(\tau) - u_0(x)}{\tau - x}\,d\tau.$$

This gives (1.89) (up to the normalizing constant that we have neglected).

Example 1.11 (Caputo derivatives and the fractional Laplacian). The time-fractional diffusion that we model by the Caputo derivative has fundamental differences with respect to the space-fractional diffusion driven by the fractional Laplacian, since the latter possesses many invariances, such as the ones under rotations and translations, that are not valid in the time-fractional setting given by the Caputo derivative, in view of a memory effect which clearly distinguishes between "the past" and "the future" and determines the "time's arrow."

On the other hand, the sum of two Caputo derivatives with opposite time directions reduces to the fractional Laplacian, as we now discuss in detail.

Let $\alpha \in (0,1)$ and let u be sufficiently smooth. An integration by parts[3] allows us to write the (left) Caputo derivative as

$$D^{\alpha}_{t,a,+}u(t) := \frac{u(t)}{\Gamma(1-\alpha)(t-a)^{\alpha}} + \frac{\alpha}{\Gamma(1-\alpha)} \int_a^t \frac{u(t)-u(\tau)}{(t-\tau)^{\alpha+1}}\,d\tau. \tag{1.92}$$

Choosing $a := -\infty$, formula (1.92) reduces to

$$D^{\alpha}_{t,-\infty,+}u(t) = \frac{\alpha}{\Gamma(1-\alpha)} \int_{-\infty}^t \frac{u(t)-u(\tau)}{(t-\tau)^{\alpha+1}}\,d\tau = \frac{\alpha}{\Gamma(1-\alpha)} \int_0^{+\infty} \frac{u(t)-u(t-\tau)}{\tau^{\alpha+1}}\,d\tau. \tag{1.93}$$

Since we will be interested in the core of this monograph in given fractional derivatives with a prescribed arrow of time, we will focus on this type of definition (and, in fact, also on higher order ones, as in (2.6)). Nevertheless, by an inversion of the time's arrow, one can also define a notion of right Caputo derivative, which, when $a := +\infty$, can be written as

$$D^{\alpha}_{t,+\infty,-}u(t) := \frac{\alpha}{\Gamma(1-\alpha)} \int_t^{+\infty} \frac{u(t)-u(\tau)}{(\tau-t)^{\alpha+1}}\,d\tau = \frac{\alpha}{\Gamma(1-\alpha)} \int_0^{+\infty} \frac{u(t)-u(t+\tau)}{\tau^{\alpha+1}}\,d\tau. \tag{1.94}$$

See the footnote 1 in Chapter 2 for further comments on the notion of left and right fractional derivatives. Summing up (1.93) and (1.94) and dropping normalization constants for simplicity, we have

$$\begin{aligned} D^{\alpha}_{t,-\infty,+}u(t) + D^{\alpha}_{t,+\infty,-}u(t) &= \int_0^{+\infty} \frac{2u(t)-u(t+\tau)-u(t-\tau)}{\tau^{\alpha+1}}\,d\tau \\ &= \frac{1}{2}\int_{-\infty}^{+\infty} \frac{2u(t)-u(t+\tau)-u(t-\tau)}{|\tau|^{\alpha+1}}\,d\tau \\ &= (-\Delta_t)^{\alpha/2}u(t), \end{aligned}$$

where we used the integral representation of the fractional Laplacian (recall (1.19)) and the obvious one-dimensional notation

$$\Delta_t := \frac{\partial^2}{\partial t^2}.$$

Therefore, the sum of left and right Caputo derivatives with initial points $-\infty$ and $+\infty$, respectively, gives, up to a multiplicative constant, the one-dimensional fractional Laplacian of fractional order $\frac{\alpha}{2}$.

Other applications of fractional equations will be discussed in Appendix A.

3 As a historical remark, we observe that formulas such as in (1.92) naturally relate different definitions of time-fractional derivatives, such as the Caputo derivative and the Marchaud fractional derivative.

2 Main results

After having discussed in detail several motivations for fractional equations in Chapter 1, we begin here the mathematically rigorous part of this monograph, and we start presenting the main original mathematical results contained in this book and their relation with the existing literature.

In this work we prove the local density of functions which annihilate a linear operator built by classical and fractional derivatives, both in space and time.

Nonlocal operators of fractional type present a variety of challenging problems in pure mathematics, also in connections with long-range phase transitions and nonlocal minimal surfaces, and are nowadays commonly exploited in a large number of models describing complex phenomena related to anomalous diffusion and boundary reactions in physics, biology, and material sciences (see, e. g., [16] for several examples, for instance in atom dislocations in crystals and water wave models). Furthermore, anomalous diffusion in the space variables can be seen as the natural counterpart of discontinuous Markov processes, thus providing important connections with problems in probability and statistics, and several applications to economy and finance (see, e. g., [47, 48] for pioneer works relating anomalous diffusion and financial models).

However, the development of time-fractional derivatives began at the end of the seventeenth century, also in view of contributions by mathematicians such as Leibniz, Euler, Laplace, Liouville, Abel, and Heaviside; see, e. g., [58, 59, 39, 60, 29] and the references therein for several interesting scientific and historical discussions. From the point of view of the applications, time-fractional derivatives naturally provide a model to comprise memory effects in the description of the phenomena under consideration.

In this work, the time-fractional derivative will be mostly described in terms of the so-called Caputo fractional derivative (see [19]), which induces a natural "direction" in the time variable, distinguishing between "past" and "future." In particular, the time direction encoded in this setting allows the analysis of "nonanticipative systems," namely, phenomena in which the state at a given time depends on past events, but not on future ones. The Caputo derivative is also related to other types of time-fractional derivatives, such as the Marchaud fractional derivative, which has applications in modeling anomalous time diffusion; see, e. g., [6, 4, 29]. See also [49, 65] for more details on fractional operators and several applications.

Here, we will take advantage of the nonlocal structure of a very general linear operator containing fractional derivatives in some variables (say, time, space, or both), in order to approximate, in the smooth sense and with arbitrary precision, any prescribed function. Remarkably, *no structural assumption* needs to be taken on the prescribed function. Therefore this approximation property reveals a *truly nonlocal behavior*, since it is in contrast with the rigidity of the functions that lie in the kernel

https://doi.org/10.1515/9783110664355-002

of classical linear operators (for instance, harmonic functions cannot approximate a function with interior maxima or minima, functions with null first derivatives are necessarily constant, and so on).

The approximation results with solutions of nonlocal operators have been first introduced in [25] for the case of the fractional Laplacian, and since then widely studied under different perspectives, including harmonic analysis; see [64, 33, 61, 62, 63]. The approximation result for the one-dimensional case of a fractional derivative of Caputo type has been considered in [15, 20], and operators involving classical time derivatives and additional classical derivatives in space have been studied in [26].

The great flexibility of solutions of fractional problems established by this type of approximation results has also consequences that go beyond the purely mathematical curiosity. For example, these results can be applied to study the evolution of biological populations, showing how a nonlocal hunting or dispersive strategy can be more convenient than one based on classical diffusion, in order to avoid waste of resources and optimize the search for food in a sparse environment; see [46, 17]. Interestingly, the theoretical descriptions provided in this setting can be compared with a series of concrete biological data and real-world experiments, confirming anomalous diffusion behaviors in many biological species; see [74].

Another interesting application of time-fractional derivatives arises in neuroscience, for instance in view of the anomalous diffusion which has been experimentally measured in neurons; see, e. g., [66] and the references therein. In this case, the anomalous diffusion could be seen as the effect of the highly ramified structure of the biological cells taken into account; see [9, 27].

In many applications, it is also natural to consider the case in which different types of diffusion take place in different variables. For instance, classical diffusion in space variables could be naturally combined to anomalous diffusion with respect to variables which take into account genetical information; see [57, 70].

Now, to state the main original results of this work, we introduce some notation. In what follows, we will denote the "local variables" with the symbol x, the "nonlocal variables" with y, and the "time-fractional variables" with t. Namely, we consider the variables

$$\begin{aligned} x &= (x_1, \dots, x_n) \in \mathbb{R}^{p_1} \times \cdots \times \mathbb{R}^{p_n}, \\ y &= (y_1, \dots, y_M) \in \mathbb{R}^{m_1} \times \cdots \times \mathbb{R}^{m_M}, \\ \text{and} \quad t &= (t_1, \dots, t_l) \in \mathbb{R}^l, \end{aligned} \tag{2.1}$$

for some $p_1, \dots, p_n, M, m_1, \dots, m_M, l \in \mathbb{N}$, and we let

$$(x, y, t) \in \mathbb{R}^N, \quad \text{where } N := p_1 + \cdots + p_n + m_1 + \cdots + m_M + l.$$

When necessary, we will use the notation B_R^k to denote the k-dimensional ball of radius R, centered at the origin in $\mathbb{R}^k$; otherwise, when there are no ambiguities, we will use the usual notation B_R.

For fixed $r = (r_1, \dots, r_n) \in \mathbb{N}^{p_1} \times \dots \times \mathbb{N}^{p_n}$, with $|r_i| \geq 1$ for each $i \in \{1, \dots, n\}$, and $\mathfrak{a} = (\mathfrak{a}_1, \dots, \mathfrak{a}_n) \in \mathbb{R}^n$, we consider the local operator acting on the variables $x = (x_1, \dots, x_n)$ given by

$$\mathfrak{l} := \sum_{i=1}^{n} \mathfrak{a}_i \partial_{x_i}^{r_i}, \tag{2.2}$$

where the multi-index notation has been used.

Furthermore, given $\mathfrak{b} = (\mathfrak{b}_1, \dots, \mathfrak{b}_M) \in \mathbb{R}^M$ and $s = (s_1, \dots, s_M) \in (0, +\infty)^M$, we consider the operator

$$\mathcal{L} := \sum_{j=1}^{M} \mathfrak{b}_j (-\Delta)_{y_j}^{s_j}, \tag{2.3}$$

where each operator $(-\Delta)_{y_j}^{s_j}$ denotes the fractional Laplacian of order $2s_j$ acting on the set of space variables $y_j \in \mathbb{R}^{m_j}$. More precisely, for any $j \in \{1, \dots, M\}$, given $s_j > 0$ and $h_j \in \mathbb{N}$ with $h_j := \min_{q_j \in \mathbb{N}}$ such that $s_j \in (0, q_j)$, in the spirit of [1], we consider the operator

$$(-\Delta)_{y_j}^{s_j} u(x, y, t) := \int_{\mathbb{R}^{m_j}} \frac{(\delta_{h_j} u)(x, y, t, Y_j)}{|Y_j|^{m_j + 2s_j}} \, dY_j, \tag{2.4}$$

where

$$(\delta_{h_j} u)(x, y, t, Y_j) := \sum_{k=-h_j}^{h_j} (-1)^k \binom{2h_j}{h_j - k} u(x, y_1, \dots, y_{j-1}, y_j + kY_j, y_{j+1}, \dots, y_M, t). \tag{2.5}$$

In particular, when $h_j := 1$, this setting comprises the case of the fractional Laplacian $(-\Delta)_{y_j}^{s_j}$ of order $2s_j \in (0, 2)$, given by

$$\begin{aligned}(-\Delta)_{y_j}^{s_j} u(x, y, t) := c_{m_j, s_j} \int_{\mathbb{R}^{m_j}} &(2u(x, y, t) - u(x, y_1, \dots, y_{j-1}, y_j + Y_j, y_{j+1}, \dots, y_M, t) \\ &- u(x, y_1, \dots, y_{j-1}, y_j - Y_j, y_{j+1}, \dots, y_M, t)) \frac{dY_j}{|Y_j|^{m_j + 2s_j}},\end{aligned}$$

where $s_j \in (0, 1)$ and c_{m_j, s_j} denotes a multiplicative normalizing constant (see, e. g., formula (3.1.10) in [16]).

It is interesting to recall that if $h_j = 2$ and $s_j = 1$ the setting in (2.4) provides a nonlocal representation for the classical Laplacian; see [4].

In our general framework, we take into account also nonlocal operators of time-fractional type. To this end, for any $\alpha > 0$, letting $k := [\alpha] + 1$ and $a \in \mathbb{R} \cup \{-\infty\}$, one can introduce the left[1] Caputo fractional derivative of order α and initial point a, defined, for $t > a$, as

$$D^{\alpha}_{t,a}u(t) := \frac{1}{\Gamma(k-\alpha)} \int_a^t \frac{\partial_t^k u(\tau)}{(t-\tau)^{\alpha-k+1}}\,d\tau, \tag{2.6}$$

where[2] Γ denotes the Euler Gamma function.

In this framework, for fixed $\mathfrak{c} = (\mathfrak{c}_1,\dots,\mathfrak{c}_l) \in \mathbb{R}^l$, $\alpha = (\alpha_1,\dots,\alpha_l) \in (0,+\infty)^l$, and $a = (a_1,\dots,a_l) \in (\mathbb{R}\cup\{-\infty\})^l$, we set

$$\mathcal{D}_a := \sum_{h=1}^{l} \mathfrak{c}_h D^{\alpha_h}_{t_h,a_h}. \tag{2.7}$$

Then, in the notation introduced in (2.2), (2.3), and (2.7), we consider here the superposition of the local, the space-fractional, and the time-fractional operators, that is, we set

$$\Lambda_a := \mathfrak{l} + \mathcal{L} + \mathcal{D}_a. \tag{2.8}$$

With this, the statement of our main result goes as follows.

Theorem 2.1. *Suppose that*

$$\begin{aligned} &\textit{either there exists } i \in \{1,\dots,M\} \textit{ such that } \mathfrak{b}_i \neq 0 \textit{ and } s_i \notin \mathbb{N}, \\ &\textit{or there exists } i \in \{1,\dots,l\} \textit{ such that } \mathfrak{c}_i \neq 0 \textit{ and } \alpha_i \notin \mathbb{N}. \end{aligned} \tag{2.9}$$

Let $\ell \in \mathbb{N}$, $f : \mathbb{R}^N \to \mathbb{R}$, with $f \in C^{\ell}(\overline{B_1^N})$. For fixed $\epsilon > 0$, there exist

$$\begin{aligned} &u = u_\epsilon \in C^\infty(B_1^N) \cap C(\mathbb{R}^N), \\ &a = (a_1,\dots,a_l) = (a_{1,\epsilon},\dots,a_{l,\epsilon}) \in (-\infty,0)^l, \\ \textit{and}\quad &R = R_\epsilon > 1 \end{aligned} \tag{2.10}$$

1 In the literature, one often finds also the notion of right Caputo fractional derivative, defined for $t < a$ by

$$\frac{(-1)^k}{\Gamma(k-\alpha)} \int_t^a \frac{\partial_t^k u(\tau)}{(\tau-t)^{\alpha-k+1}}\,d\tau.$$

Since the right time-fractional derivative boils down to the left one (by replacing t with $2a-t$), in this work we focus only on the case of left derivatives.

Also, though there are several time-fractional derivatives that are studied in the literature under different perspectives, we focus here on the Caputo derivative, since it possesses well-posedness properties with respect to classical initial value problems, differently than other time-fractional derivatives, such as the Riemann–Liouville derivative, in which the initial value setting involves data containing derivatives of fractional order.

2 For notational simplicity, we will often denote $\partial_t^k u = u^{(k)}$.

such that

$$\begin{cases} \Lambda_a u = 0 & \text{in } B_1^N, \\ u = 0 & \text{in } \mathbb{R}^N \setminus B_R^N, \end{cases} \tag{2.11}$$

and

$$\|u - f\|_{C^\ell(B_1^N)} < \epsilon. \tag{2.12}$$

We observe that the initial points of the Caputo-type operators in Theorem 2.1 also depend on ϵ, as detailed in (2.10) (but the other parameters, such as the orders of the operators involved, are fixed arbitrarily).

We also stress that condition (2.9) requires that the operator Λ_a contains at least one nonlocal operator among its building blocks in (2.2), (2.3), and (2.7). This condition cannot be avoided, since approximation results in the same spirit of Theorem 2.1 cannot hold for classical differential operators.

Theorem 2.1 comprises, as particular cases, the nonlocal approximation results established in the recent literature of this topic. Indeed, when

$$\begin{aligned} &\mathfrak{a}_1 = \cdots = \mathfrak{a}_n = \mathfrak{b}_1 = \cdots = \mathfrak{b}_{M-1} = \mathfrak{c}_1 = \cdots = \mathfrak{c}_l = 0, \\ &\mathfrak{b}_M = 1, \\ \text{and} \quad &s \in (0,1), \end{aligned}$$

we see that Theorem 2.1 recovers the main result in [25], giving the local density of s-harmonic functions vanishing outside a compact set.

Similarly, when

$$\begin{aligned} &\mathfrak{a}_1 = \cdots = \mathfrak{a}_n = \mathfrak{b}_1 = \cdots = \mathfrak{b}_M = \mathfrak{c}_1 = \cdots = \mathfrak{c}_{l-1} = 0, \\ &\mathfrak{c}_l = 1, \\ \text{and} \quad &\mathcal{D}_a = D^\alpha_{t,a}, \quad \text{for some } \alpha > 0,\, a < 0, \end{aligned}$$

Theorem 2.1 reduces to the main results in [15] for $\alpha \in (0,1)$ and [20] for $\alpha > 1$, in which such approximation result was established for Caputo-stationary functions, i. e., functions that annihilate the Caputo fractional derivative.

Also, when

$$\begin{aligned} &p_1 = \cdots = p_n = 1, \\ &\mathfrak{c}_1 = \cdots = \mathfrak{c}_l = 0, \\ \text{and} \quad &s_j \in (0,1), \quad \text{for every } j \in \{1,\ldots,M\}, \end{aligned}$$

Theorem 2.1 recovers the cases taken into account in [26], in which approximation results have been established for the superposition of a local operator with a superposition of fractional Laplacians of order $2s_j < 2$.

In this sense, not only Theorem 2.1 comprises the existing literature, but it goes beyond it, since it combines classical derivatives, fractional Laplacians, and Caputo fractional derivatives altogether. In addition, it comprises the cases in which the space-fractional Laplacians taken into account are of order greater than 2.

As a matter of fact, this point is also a novelty introduced by Theorem 2.1 here with respect to the previous literature.

Theorem 2.1 was announced in [20], and we have just received the very interesting preprint [42] which also considered the case of different, not necessarily fractional, powers of the Laplacian, using a different and innovative methodology.

The rest of this book is organized as follows. Chapter 3 focuses on time-fractional operators. More precisely, in Sections 3.1 and 3.2 we study the boundary behavior of the eigenfunctions of the Caputo derivative and of functions with vanishing Caputo derivative, respectively, detecting their singular boundary behavior in terms of explicit representation formulas. These type of results are interesting in themselves and can also find further applications.

Chapter 4 is devoted to some properties of the higher order fractional Laplacian. More precisely, Section 4.1 provides some representation formula of the solution of $(-\Delta)^s u = f$ in a ball, with $u = 0$ outside this ball, for all $s > 0$, and extends the Green formula methods introduced in [24] and [2].

Then, in Section 4.2 we study the boundary behavior of the first Dirichlet eigenfunction of higher order fractional equations, and in Section 4.3 we give some precise asymptotics at the boundary for the first Dirichlet eigenfunction of $(-\Delta)^s$ for any $s > 0$.

Section 4.4 is devoted to the analysis of the asymptotic behavior of s-harmonic functions, with a “spherical bump function” as exterior Dirichlet datum.

Chapter 5 is devoted to the proof of our main result. To this end, Section 5.1 contains an auxiliary statement, namely, Theorem 5.1, which will imply Theorem 2.1. This is technically convenient, since the operator Λ_a depends in principle on the initial point a. This has the disadvantage that if $\Lambda_a u_a = 0$ and $\Lambda_b u_b = 0$ in some domain, the function $u_a + u_b$ is not in principle a solution of any operator, unless $a = b$. To overcome such a difficulty, in Theorem 5.1 we will restrict ourselves to the case in which $a = -\infty$, exploiting a polynomial extension that we have introduced and used in [20].

In Section 5.2 we make the main step towards the proof of Theorem 5.1. Here, we prove that functions in the kernel of nonlocal operators such as the one in (2.8) span with their derivatives a maximal Euclidean space. This fact is special for the nonlocal case and its proof is based on the boundary analysis of the fractional operators in both time and space. Due to the general form of the operator in (2.8), we have to distinguish here several cases, taking advantage of either the time-fractional or the space-fractional components of the operators.

Finally, in Section 5.3 we complete the proof of Theorem 5.1, using the previous approximation results and suitable rescaling arguments.

The final appendix provides concrete examples in which our main result can be applied.

3 Boundary behavior of solutions of time-fractional equations

In this chapter, we give precise asymptotics for the boundary behavior of solutions of time-fractional equations. The cases of the eigenfunctions and of the Dirichlet problem with vanishing forcing term will be studied in detail (the latter will be often referred to as the time-fractional harmonic case, borrowing terminology from elliptic equations, with a slight abuse of notation in our case).

3.1 Sharp boundary behavior for the time-fractional eigenfunctions

In this section we show that the eigenfunctions of the Caputo fractional derivative in (2.6) have an explicit representation via the Mittag-Leffler function. For this, for fixed $\alpha, \beta \in \mathbb{C}$ with $\mathbb{R}(\alpha) > 0$, for any z with $\mathbb{R}(z) > 0$, we recall that the Mittag-Leffler function is defined as

$$E_{\alpha,\beta}(z) := \sum_{j=0}^{+\infty} \frac{z^j}{\Gamma(\alpha j + \beta)}. \tag{3.1}$$

The Mittag-Leffler function plays an important role in equations driven by the Caputo derivatives, replacing the exponential function for classical differential equations, as given by the following well-established result (see [36] and the references therein).

Lemma 3.1. *Let $\alpha \in (0,1]$, $\lambda \in \mathbb{R}$, and $a \in \mathbb{R} \cup \{-\infty\}$. Then the unique solution of the boundary value problem*

$$\begin{cases} D^{\alpha}_{t,a} u(t) = \lambda u(t) & \text{for any } t \in (a,+\infty), \\ u(a) = 1 \end{cases}$$

is given by $E_{\alpha,1}(\lambda(t-a)^{\alpha})$.

Lemma 3.1 can be actually generalized[1] to any fractional order of differentiation α.

Lemma 3.2. *Let $\alpha \in (0,+\infty)$, with $\alpha \in (k-1,k]$, $k \in \mathbb{N}$, $a \in \mathbb{R} \cup \{-\infty\}$, and $\lambda \in \mathbb{R}$. Then the unique continuous solution of the boundary value problem*

$$\begin{cases} D^{\alpha}_{t,a} u(t) = \lambda u(t) & \text{for any } t \in (a,+\infty), \\ u(a) = 1, \\ \partial^m_t u(a) = 0 & \text{for any } m \in \{1,\dots,k-1\} \end{cases} \tag{3.2}$$

is given by $u(t) = E_{\alpha,1}(\lambda(t-a)^{\alpha})$.

1 It is easily seen that for $k := 1$ Lemma 3.2 boils down to Lemma 3.1.

https://doi.org/10.1515/9783110664355-003

Proof. For the sake of simplicity we take $a = 0$. Also, the case in which $\alpha \in \mathbb{N}$ can be checked with a direct computation, so we focus on the case $\alpha \in (k-1, k)$, with $k \in \mathbb{N}$.

We let $u(t) := E_{\alpha,1}(\lambda t^\alpha)$. It is straightforward to see that $u(t) = 1 + \mathcal{O}(t^k)$ and therefore

$$u(0) = 1 \quad \text{and} \quad \partial_t^m u(0) = 0 \quad \text{for any } m \in \{1, \dots, k-1\}. \tag{3.3}$$

We also claim that

$$D_{t,a}^\alpha u(t) = \lambda u(t) \quad \text{for any } t \in (0, +\infty). \tag{3.4}$$

To check this, we recall (2.6) and (3.1) (with $\beta := 1$), and we have

$$\begin{aligned}
&D_{t,a}^\alpha u(t)\\
&= \frac{1}{\Gamma(k-\alpha)} \int_0^t \frac{u^{(k)}(\tau)}{(t-\tau)^{\alpha-k+1}}\, d\tau\\
&= \frac{1}{\Gamma(k-\alpha)} \int_0^t \left(\sum_{j=1}^{+\infty} \lambda^j \frac{\alpha j(\alpha j-1)\cdots(\alpha j-k+1)}{\Gamma(\alpha j+1)} \tau^{\alpha j-k} \right) \frac{d\tau}{(t-\tau)^{\alpha-k+1}}\\
&= \sum_{j=1}^{+\infty} \lambda^j \frac{\alpha j(\alpha j-1)\cdots(\alpha j-k+1)}{\Gamma(k-\alpha)\Gamma(\alpha j+1)} \int_0^t \tau^{\alpha j-k}(t-\tau)^{k-\alpha-1}\, d\tau.
\end{aligned}$$

Hence, using the change of variable $\tau = t\sigma$, we obtain

$$D_{t,a}^\alpha u(t) = \sum_{j=1}^{+\infty} \lambda^j \frac{\alpha j(\alpha j-1)\cdots(\alpha j-k+1)}{\Gamma(k-\alpha)\Gamma(\alpha j+1)} t^{\alpha j-\alpha} \int_0^1 \sigma^{\alpha j-k}(1-\sigma)^{k-\alpha-1}\, d\tau. \tag{3.5}$$

However, from the basic properties of the Beta function, it is known that if $\Re(z)$, $\Re(w) > 0$, then

$$\int_0^1 \sigma^{z-1}(1-\sigma)^{w-1}\, dt = \frac{\Gamma(z)\Gamma(w)}{\Gamma(z+w)}. \tag{3.6}$$

In particular, taking $z := \alpha j - k + 1 \in (\alpha - k + 1, +\infty) \subseteq (0, +\infty)$ and $w := k - \alpha \in (0, +\infty)$, and substituting (3.6) into (3.5), we conclude that

$$\begin{aligned}
D_{t,a}^\alpha u(t) &= \sum_{j=1}^{+\infty} \lambda^j \frac{\alpha j(\alpha j-1)\cdots(\alpha j-k+1)}{\Gamma(k-\alpha)\Gamma(\alpha j+1)} \frac{\Gamma(\alpha j-k+1)\Gamma(k-\alpha)}{\Gamma(\alpha j-\alpha+1)} t^{\alpha j-\alpha}\\
&= \sum_{j=1}^{+\infty} \lambda^j \frac{\alpha j(\alpha j-1)\cdots(\alpha j-k+1)}{\Gamma(\alpha j+1)} \frac{\Gamma(\alpha j-k+1)}{\Gamma(\alpha j-\alpha+1)} t^{\alpha j-\alpha}.
\end{aligned} \tag{3.7}$$

Now we use the fact that $z\Gamma(z) = \Gamma(z+1)$ for any $z \in \mathbb{C}$ with $\Re(z) > -1$, so we have

$$\alpha j(\alpha j-1)\cdots(\alpha j-k+1)\Gamma(\alpha j-k+1) = \Gamma(\alpha j+1).$$

Plugging this information into (3.7), we find

$$D^{\alpha}_{t,a}u(t) = \sum_{j=1}^{+\infty} \frac{\lambda^j}{\Gamma(\alpha j - \alpha + 1)} t^{\alpha j - \alpha} = \sum_{j=0}^{+\infty} \frac{\lambda^{j+1}}{\Gamma(\alpha j + 1)} t^{\alpha j} = \lambda u(t).$$

This proves (3.4).

Then, in view of (3.3) and (3.4) we obtain that u is a solution of (3.2). Hence, to complete the proof of the desired result, we have to show that such a solution is unique. To this end, supposing that we have two solutions of (3.2), we consider their difference w, and we observe that w is a solution of

$$\begin{cases} D^{\alpha}_{t,0}w(t) = \lambda w(t) & \text{for any } t \in (0, +\infty), \\ \partial^m_t w(0) = 0 & \text{for any } m \in \{0, \dots, k-1\}. \end{cases}$$

By Theorem 4.1 in [69], it follows that w vanishes identically, and this proves the desired uniqueness result. □

The boundary behavior of the Mittag-Leffler function for different values of the fractional parameter α is depicted in Figure 3.1. In light of (3.1), we note in particular that, near $z = 0$,

$$E_{\alpha,\beta}(z) = \frac{1}{\Gamma(\beta)} + \frac{z}{\Gamma(\alpha+\beta)} + O(z^2)$$

and therefore, near $t = a$,

$$E_{\alpha,1}(\lambda(t-a)^{\alpha}) = 1 + \frac{\lambda(t-a)^{\alpha}}{\Gamma(\alpha+1)} + O(\lambda^2(t-a)^{2\alpha}).$$

3.2 Sharp boundary behavior for the time-fractional harmonic functions

In this section, we detect the optimal boundary behavior of time-fractional harmonic functions and of their derivatives. The result that we need for our purposes is the following.

Lemma 3.3. *Let $\alpha \in (0, +\infty) \setminus \mathbb{N}$. There exists a function $\psi : \mathbb{R} \to \mathbb{R}$ such that $\psi \in C^{\infty}((1, +\infty))$ and*

$$D^{\alpha}_0 \psi(t) = 0 \quad \text{for all } t \in (1, +\infty), \tag{3.8}$$

$$\text{and} \quad \lim_{\epsilon \searrow 0} \epsilon^{\ell-\alpha} \partial^{\ell} \psi(1 + \epsilon t) = \kappa_{\alpha,\ell} t^{\alpha-\ell}, \quad \text{for all } \ell \in \mathbb{N}, \tag{3.9}$$

for some $\kappa_{\alpha,\ell} \in \mathbb{R} \setminus \{0\}$, where (3.9) is taken in the sense of distribution for $t \in (0, +\infty)$.

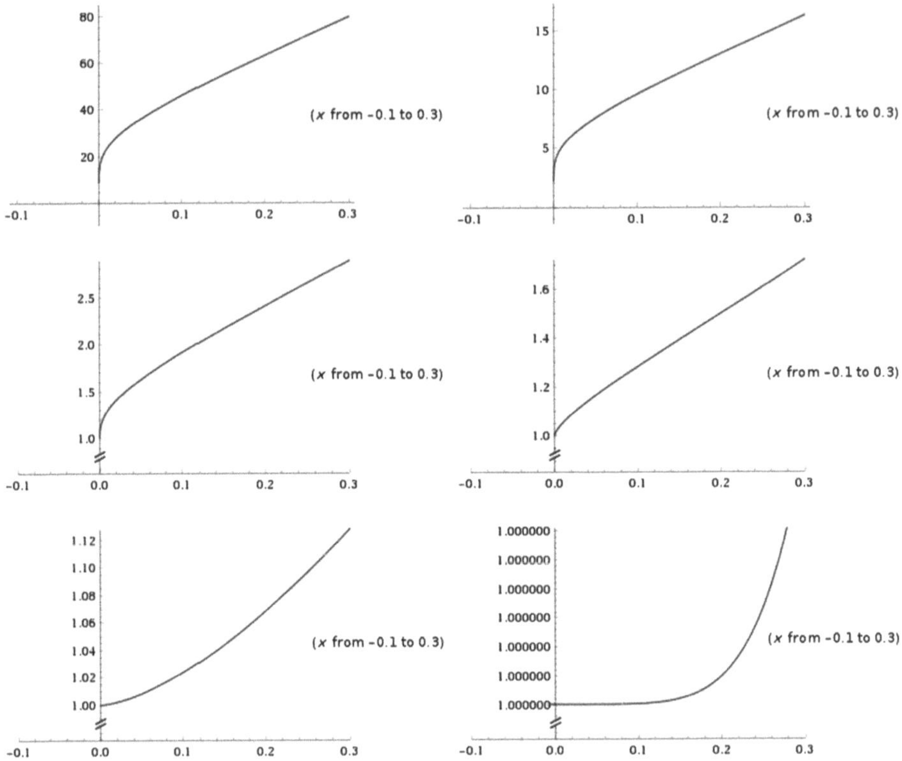

Figure 3.1: Behavior of the Mittag-Leffler function $E_{\alpha,1}(t^{\alpha})$ near the origin for $\alpha = \frac{1}{100}$, $\alpha = \frac{1}{20}$, $\alpha = \frac{1}{3}$, $\alpha = \frac{2}{3}$, $\alpha = \frac{3}{2}$, and $\alpha = \frac{11}{2}$.

Proof. We use Lemma 2.5 in [20], according to which (see in particular formula (2.16) in [20]) the claim in (3.8) holds. Furthermore (see formulas (2.19) and (2.20) in [20]), we can write that, for all $t > 1$,

$$\psi(t) = -\frac{1}{\Gamma(\alpha)\Gamma([\alpha]+1-\alpha)} \iint\limits_{[1,t]\times[0,3/4]} \partial^{[\alpha]+1}\psi_0(\sigma)(\tau-\sigma)^{[\alpha]-\alpha}(t-\tau)^{\alpha-1}\,d\tau\,d\sigma, \tag{3.10}$$

for a suitable $\psi_0 \in C^{[\alpha]+1}([0,1])$.

In addition, by Lemma 2.6 in [20], we can write

$$\lim_{\epsilon\searrow 0} \epsilon^{-\alpha}\psi(1+\epsilon) = \kappa, \tag{3.11}$$

for some $\kappa \neq 0$. Now we set

$$(0,+\infty) \ni t \mapsto f_{\epsilon}(t) := \epsilon^{\ell-\alpha}\partial^{\ell}\psi(1+\epsilon t).$$

We observe that, for any $\varphi \in C_0^\infty((0,+\infty))$,

$$\int_0^{+\infty} f_\epsilon(t)\varphi(t)\,dt = \epsilon^{\ell-\alpha}\int_0^{+\infty} \partial^\ell \psi(1+\epsilon t)\varphi(t)\,dt$$

$$= \epsilon^{-\alpha}\int_0^{+\infty} \frac{d^\ell}{dt^\ell}(\psi(1+\epsilon t))\varphi(t)\,dt = (-1)^\ell \epsilon^{-\alpha}\int_0^{+\infty} \psi(1+\epsilon t)\partial^\ell\varphi(t)\,dt. \quad (3.12)$$

Also, in view of (3.10),

$$\begin{aligned}
&\epsilon^{-\alpha}|\psi(1+\epsilon t)| \\
&= \left|\frac{\epsilon^{-\alpha}}{\Gamma(\alpha)\Gamma([\alpha]+1-\alpha)}\iint_{[1,1+\epsilon t]\times[0,3/4]} \partial^{[\alpha]+1}\psi_0(\sigma)(\tau-\sigma)^{[\alpha]-\alpha}(1+\epsilon t-\tau)^{\alpha-1}\,d\tau\,d\sigma\right| \\
&\le C\epsilon^{-\alpha}\int_{[1,1+\epsilon t]} (1+\epsilon t-\tau)^{\alpha-1}\,d\tau \\
&= Ct^\alpha,
\end{aligned}$$

which is locally bounded in t, where $C > 0$ here above may vary from line to line.

As a consequence, we can pass to the limit in (3.12) and obtain

$$\lim_{\epsilon\searrow 0}\int_0^{+\infty} f_\epsilon(t)\varphi(t)\,dt = (-1)^\ell \int_0^{+\infty} \lim_{\epsilon\searrow 0}\epsilon^{-\alpha}\psi(1+\epsilon t)\partial^\ell\varphi(t)\,dt.$$

This and (3.11) give

$$\lim_{\epsilon\searrow 0}\int_0^{+\infty} f_\epsilon(t)\varphi(t)\,dt = (-1)^\ell\kappa\int_0^{+\infty} t^\alpha\partial^\ell\varphi(t)\,dt = \kappa\alpha\cdots(\alpha-\ell+1)\int_0^{+\infty} t^{\alpha-\ell}\varphi(t)\,dt,$$

which establishes (3.9). □

4 Boundary behavior of solutions of space-fractional equations

In this chapter, we give precise asymptotics for the boundary behavior of solutions of space-fractional equations. The cases of the eigenfunctions and of the Dirichlet problem with vanishing forcing term will be studied in detail. To this end, we will also exploit useful representation formulas of the solutions in terms of suitable Green functions.

4.1 Green representation formulas and solution of $(-\Delta)^s u = f$ in B_1 with homogeneous Dirichlet datum

Our goal is to provide some representation results on the solution of $(-\Delta)^s u = f$ in a ball, with $u = 0$ outside this ball, for all $s > 0$. Our approach is an extension of the Green formula methods introduced in [24] and [2] (see also [14] for the case $s \in (0,1)$). Differently from the previous literature, we are not assuming here that f is regular in the whole of the ball, but merely that it is Hölder continuous near the boundary and sufficiently integrable inside. Given the type of singularity of the Green function, these assumptions are sufficient to obtain meaningful representations, which in turn will be useful to deal with the eigenfunction problem in Section 4.2.

4.1.1 Solving $(-\Delta)^s u = f$ in B_1 for discontinuous f vanishing near ∂B_1

Now, we want to extend the representation results of [24] and [2] to the case in which the right hand side is not Hölder continuous, but merely in a Lebesgue space, but it has the additional property of vanishing near the boundary of the domain. To this end, for fixed $s > 0$, we consider the polyharmonic Green function in $B_1 \subset \mathbb{R}^n$, given, for every $x \neq y \in \mathbb{R}^n$, by

$$\begin{aligned} \mathcal{G}_s(x,y) &:= \frac{k(n,s)}{|x-y|^{n-2s}} \int_0^{r_0(x,y)} \frac{\eta^{s-1}}{(\eta+1)^{\frac{n}{2}}}\, d\eta, \\ \text{where} \quad r_0(x,y) &:= \frac{(1-|x|^2)_+(1-|y|^2)_+}{|x-y|^2}, \\ \text{with} \quad k(n,s) &:= \frac{\Gamma(\frac{n}{2})}{\pi^{\frac{n}{2}} 4^s \Gamma^2(s)}. \end{aligned} \tag{4.1}$$

Given $x \in B_1$, we also set

$$d(x) := 1 - |x|. \tag{4.2}$$

In this setting, we have the following.

https://doi.org/10.1515/9783110664355-004

Proposition 4.1. *Let $r \in (0,1)$ and $f \in L^2(B_1)$, with $f = 0$ in $\mathbb{R}^n \setminus B_r$. Let*

$$u(x) := \begin{cases} \int_{B_1} \mathcal{G}_s(x,y) f(y)\, dy & \text{if } x \in B_1, \\ 0 & \text{if } x \in \mathbb{R}^n \setminus B_1. \end{cases} \tag{4.3}$$

Then

$$u \in L^1(B_1), \quad \text{and} \quad \|u\|_{L^1(B_1)} \le C\|f\|_{L^1(B_1)}, \tag{4.4}$$

$$\text{for every } R \in (r,1), \qquad \sup_{x \in B_1 \setminus B_R} d^{-s}(x)|u(x)| \le C_R \|f\|_{L^1(B_1)}, \tag{4.5}$$

$$u \text{ satisfies } (-\Delta)^s u = f \text{ in } B_1 \text{ in the sense of distributions,} \tag{4.6}$$

and

$$u \in W^{2s,2}_{loc}(B_1). \tag{4.7}$$

Here, $C > 0$ is a constant depending on n, s, and r, $C_R > 0$ is a constant depending on n, s, r, and R, and $C_\rho > 0$ is a constant depending on n, s, r, and ρ.

When $f \in C^\alpha(B_1)$ for some $\alpha \in (0,1)$, Proposition 4.1 boils down to the main results of [24] and [2].

Proof of Proposition 4.1. We recall the following useful estimate; see Lemma 3.3 in [2]. For any $\epsilon \in (0, \min\{n,s\})$ and any $\bar{R}, \bar{r} > 0$,

$$\frac{1}{\bar{R}^{n-2s}} \int_0^{\bar{r}/\bar{R}^2} \frac{\eta^{s-1}}{(\eta+1)^{\frac{n}{2}}}\, d\eta \le \frac{2}{s} \frac{\bar{r}^{s-(\epsilon/2)}}{\bar{R}^{n-\epsilon}},$$

and so, by (4.1) and (4.2), for every $x, y \in B_1$,

$$\mathcal{G}_s(x,y) \le \frac{C d^{s-(\epsilon/2)}(x) d^{s-(\epsilon/2)}(y)}{|x-y|^{n-\epsilon}}$$

for some $C > 0$. Hence, recalling (4.3),

$$\begin{aligned} \int_{B_1} |u(x)|\, dx &\le \int_{B_1} \left(\int_{B_1} \mathcal{G}_s(x,y)|f(y)|\, dy \right) dx \\ &\le C \int_{B_1} \left(\int_{B_1} \frac{|f(y)|}{|x-y|^{n-\epsilon}}\, dy \right) dx \\ &= C \int_{B_1} \left(\int_{B_1} \frac{|f(y)|}{|x-y|^{n-\epsilon}}\, dx \right) dy \\ &= C \int_{B_1} |f(y)|\, dy, \end{aligned}$$

up to renaming $C > 0$ line after line, and this proves (4.4).

Now, if $x \in B_1 \setminus B_R$ and $y \in B_r$, with $0 < r < R < 1$, we have

$$|x-y| \geq |x| - |y| \geq R - r$$

and accordingly

$$r_0(x,y) \leq \frac{2d(x)}{(R-r)^2},$$

which in turn implies that

$$\mathcal{G}_s(x,y) \leq \frac{k(n,s)}{|x-y|^{n-2s}} \int_0^{2d(x)/(R-r)^2} \frac{\eta^{s-1}}{(\eta+1)^{\frac{n}{2}}}\, d\eta \leq C d^s(x),$$

for some $C > 0$. As a consequence, since f vanishes outside B_r, we see that, for any $x \in B_1 \setminus B_R$,

$$|u(x)| \leq \int_{B_r} \mathcal{G}_s(x,y)|f(y)|\, dy \leq C d^s(x) \int_{B_r} |f(y)|\, dy,$$

which proves (4.5).

Now, we fix $\hat{r} \in (r,1)$ and consider a mollification of f, which we denote by $f_j \in C_0^\infty(B_{\hat{r}})$, with $f_j \to f$ in $L^2(B_1)$ as $j \to +\infty$. We also write $\mathcal{G}_s * f$ as a short notation for the right hand side of (4.3). Then, by [24] and [2], we know that $u_j := \mathcal{G}_s * f_j$ is a (locally smooth, hence distributional) solution of $(-\Delta)^s u_j = f_j$. Furthermore, if we set $\tilde{u}_j := u_j - u$ and $\tilde{f}_j := f_j - f$ we have

$$\tilde{u}_j = \mathcal{G}_s * (f_j - f) = \mathcal{G}_s * \tilde{f}_j,$$

and therefore, by (4.4),

$$\|\tilde{u}_j\|_{L^1(B_1)} \leq C \|\tilde{f}_j\|_{L^1(B_1)},$$

which is infinitesimal as $j \to +\infty$. It follows that $u_j \to u$ in $L^1(B_1)$ as $j \to +\infty$, and consequently, for any $\varphi \in C_0^\infty(B_1)$,

$$\begin{aligned}\int_{B_1} u(x)(-\Delta)^s \varphi(x)\, dx &= \lim_{j \to +\infty} \int_{B_1} u_j(x)(-\Delta)^s \varphi(x)\, dx \\ &= \lim_{j \to +\infty} \int_{B_1} f_j(x)\varphi(x)\, dx = \int_{B_1} f(x)\varphi(x)\, dx,\end{aligned}$$

thus completing the proof of (4.6).

Now, to prove (4.7), we suppose that $s \in (0,+\infty) \setminus \mathbb{N}$, since the case of integer s is classical; see, e. g., [34]. First of all, we claim that

$$\text{(4.7) holds for every } s \in (0,1). \tag{4.8}$$

For this, we first claim that if $g \in C^\infty(B_1)$ and v is a (locally smooth) solution of $(-\Delta)^s v = g$ in B_1, with $v = 0$ outside B_1, then $v \in W^{2s,2}_{loc}(B_1)$, and, for any $\rho \in (0, 1)$,

$$\|v\|_{W^{2s,2}(B_\rho)} \leq C_\rho \|g\|_{L^2(B_1)}. \tag{4.9}$$

This claim can be seen as a localization of Lemma 3.1 of [28], or a quantification of the last claim in Theorem 1.3 of [12]. To prove (4.9), we let $R_- < R_+ \in (\rho, 1)$, and we consider $\eta \in C_0^\infty(B_{R_+})$ with $\eta = 1$ in B_{R_-}. We let $v^* := v\eta$, and we recall formulas (3.2), (3.3), and (A.5) in [12], according to which

$$\begin{aligned} &(-\Delta)^s v^* - \eta(-\Delta)^s v = g^* \quad \text{in } \mathbb{R}^n, \\ \text{with} \quad &\|g^*\|_{L^2(\mathbb{R}^n)} \leq C\|v\|_{W^{s,2}(\mathbb{R}^n)}, \end{aligned}$$

for some $C > 0$.

Moreover, using a notation taken from [12] we denote by $W_0^{s,2}(\overline{B_1})$ the space of functions in $W^{s,2}(\mathbb{R}^n)$ vanishing outside B_1 and we consider the dual space $W_0^{-s,2}(\overline{B_1})$. We remark that if $h \in L^2(B_1)$ we can naturally identify h as an element of $W_0^{-s,2}(\overline{B_1})$ by considering the action of h on any $\varphi \in W_0^{s,2}(\overline{B_1})$ as defined by

$$\int_{B_1} h(x)\varphi(x)\,dx.$$

With respect to this, we have

$$\|h\|_{W_0^{-s,2}(\overline{B_1})} = \sup_{\substack{\varphi \in W_0^{s,2}(\overline{B_1}) \\ \|\varphi\|_{W_0^{s,2}(\overline{B_1})}=1}} \int_{B_1} h(x)\varphi(x)\,dx \leq \|h\|_{L^2(B_1)}. \tag{4.10}$$

We also note that

$$\|v\|_{W^{s,2}(\mathbb{R}^n)} \leq C\|g\|_{W^{-s,2}(\overline{B_1})},$$

in light of Proposition 2.1 of [12]. This and (4.10) give

$$\|v\|_{W^{s,2}(\mathbb{R}^n)} \leq C\|g\|_{L^2(B_1)}.$$

Then, by Lemma 3.1 of [28] (see in particular formula (3.2) there, applied here with $\lambda := 0$), we obtain

$$\begin{aligned} \|v^*\|_{W^{2s,2}(\mathbb{R}^n)} &\leq C\|\eta(-\Delta)^s v + g^*\|_{L^2(\mathbb{R}^n)} \\ &\leq C(\|(-\Delta)^s v\|_{L^2(B_{R_+})} + \|g^*\|_{L^2(\mathbb{R}^n)}) \\ &= C(\|g\|_{L^2(B_{R_+})} + \|g^*\|_{L^2(\mathbb{R}^n)}) \\ &\leq C(\|g\|_{L^2(B_1)} + \|v\|_{W^{s,2}(\mathbb{R}^n)}) \\ &\leq C\|g\|_{L^2(B_1)}, \end{aligned} \tag{4.11}$$

up to renaming $C > 0$ step by step. However, since $v^* = v$ in B_ρ,

$$\|v\|_{W^{2s,2}(B_\rho)} = \|v^*\|_{W^{2s,2}(B_\rho)} \le \|v^*\|_{W^{2s,2}(\mathbb{R}^n)}.$$

From this and (4.11), we obtain (4.9), as desired.

Now, we let f_j, $\tilde{f}_j$, u_j, and $\tilde{u}_j$ as above and make use of (4.9) to write

$$\|u_j\|_{W^{2s,2}(B_\rho)} \le C_\rho \|f_j\|_{L^2(B_1)}$$
$$\text{and} \quad \|\tilde{u}_j\|_{W^{2s,2}(B_\rho)} \le C_\rho \|\tilde{f}_j\|_{L^2(B_1)}. \tag{4.12}$$

As a consequence, taking the limit as $j \to +\infty$ we obtain

$$\|u\|_{W^{2s,2}(B_\rho)} \le C_\rho \|f\|_{L^2(B_1)},$$

that is (4.7) in this case, namely, the claim in (4.8).

Now, to prove (4.7), we argue by induction on the integer part of s. When the integer part of s is zero, the basis of the induction is warranted by (4.8). Then, to perform the inductive step, given $s \in (0, +\infty) \setminus \mathbb{N}$, we suppose that (4.7) holds for $s-1$, namely,

$$\mathcal{G}_{s-1} * f \in W^{2s-2,2}_{loc}(B_1). \tag{4.13}$$

Then, following [2], it is convenient to introduce the notation

$$[x, y] := \sqrt{|x|^2|y|^2 - 2x \cdot y + 1}$$

and consider the auxiliary kernel given, for every $x \ne y \in B_1$, by

$$P_{s-1}(x,y) := \frac{(1-|x|^2)_+^{s-2}(1-|y|^2)_+^{s-1}(1-|x|^2|y|^2)}{[x,y]^n}. \tag{4.14}$$

We point out that if $x \in B_r$ with $r \in (0,1)$, then

$$[x,y]^2 \ge |x|^2|y|^2 - 2|x||y| + 1 = (1-|x||y|)^2 \ge (1-r)^2 > 0. \tag{4.15}$$

Consequently, since f is supported in B_r,

$$P_{s-1} * f \in C^\infty(\mathbb{R}^n). \tag{4.16}$$

Then, we recall that

$$-\Delta_x \mathcal{G}_s(x,y) = \mathcal{G}_{s-1}(x,y) - CP_{s-1}(x,y), \tag{4.17}$$

for some $C \in \mathbb{R}$; see Lemma 3.1 in [2].

As a consequence, in view of (4.13), (4.16), (4.17), we conclude that

$$-\Delta(\mathcal{G}_s * f) = (-\Delta_x \mathcal{G}_s) * f \in W^{2s-2,2}_{loc}(B_1).$$

This and the classical elliptic regularity theory (see, e. g., [34]) give $\mathcal{G}_s * f \in W^{2s,2}_{loc}(B_1)$, which completes the inductive proof and establishes (4.7). □

4.1.2 Solving $(-\Delta)^s u = f$ in B_1 for f Hölder continuous near ∂B_1

Our goal is now to extend the representation results of [24] and [2] to the case in which the right hand side is not Hölder continuous in the whole ball, but merely in a neighborhood of the boundary. This result is obtained here by superposing the results in [24] and [2] with Proposition 4.1 here, taking advantage of the linear structure of the problem.

Proposition 4.2. *Let $f \in L^2(B_1)$. Let $\alpha, r \in (0,1)$ and assume that*

$$f \in C^\alpha(B_1 \setminus B_r). \tag{4.18}$$

In the notation of (4.1), *let*

$$u(x) := \begin{cases} \int_{B_1} \mathcal{G}_s(x,y) f(y)\, dy & \text{if } x \in B_1, \\ 0 & \text{if } x \in \mathbb{R}^n \setminus B_1. \end{cases} \tag{4.19}$$

Then, in the notation of (4.2), *we have the following:*

$$\text{for every } R \in (r,1), \quad \sup_{x\in B_1\setminus B_R} d^{-s}(x)|u(x)| \le C_R(\|f\|_{L^1(B_1)} + \|f\|_{L^\infty(B_1\setminus B_r)}), \tag{4.20}$$

$$u \text{ satisfies } (-\Delta)^s u = f \text{ in } B_1 \text{ in the sense of distributions,} \tag{4.21}$$

and

$$u \in W^{2s,2}_{loc}(B_1). \tag{4.22}$$

Here, $C > 0$ is a constant depending on n, s, and r, $C_R > 0$ is a constant depending on n, s, r, and R, and $C_\rho > 0$ is a constant depending on n, s, r, and ρ.

Proof. We take $r_1 \in (r,1)$ and $\eta \in C_0^\infty(B_{r_1})$ with $\eta = 1$ in B_r. Let also

$$f_1 := f\eta \quad \text{and} \quad f_2 := f - f_1.$$

We observe that $f_1 \in L^2(B_1)$ and that $f_1 = 0$ outside B_{r_1}. Therefore, we are in the position of applying Proposition 4.1 and we find a function u_1 (obtained by convolving $\mathcal{G}_s$ against f_1) such that

$$\text{for every } R \in (r_1,1), \quad \sup_{x\in B_1\setminus B_R} d^{-s}(x)|u_1(x)| \le C_R\|f_1\|_{L^1(B_1)}, \tag{4.23}$$

$$u_1 \text{ satisfies } (-\Delta)^s u_1 = f_1 \text{ in } B_1 \text{ in the sense of distributions,} \tag{4.24}$$

$$\text{and} \quad u_1 \in W^{2s,2}_{loc}(B_1). \tag{4.25}$$

However, $f_2 = f(1-\eta)$ vanishes outside $B_1 \setminus B_r$ and it is Hölder continuous. Accordingly, we can apply Theorem 1.1 of [2] and find a function u_2 (obtained by convolving $\mathcal{G}_s$ against f_2) such that

$$\text{for every } R \in (r_1, 1), \qquad \sup_{x \in B_1 \setminus B_R} d^{-s}(x)|u_2(x)| \leq C_R \|f_2\|_{L^\infty(B_1)}, \tag{4.26}$$

$$u_2 \text{ satisfies } (-\Delta)^s u_2 = f_2 \text{ in } B_1 \text{ in the sense of distributions,} \tag{4.27}$$

$$\text{and} \quad u_2 \in C^{2s+\alpha}_{loc}(B_1). \tag{4.28}$$

Then, $f = f_1 + f_2$, and thus, in view of (4.19), we have $u = u_1 + u_2$. Also, u satisfies (4.20) thanks to (4.23) and (4.26), (4.21) thanks to (4.24) and (4.27), and (4.22) thanks to (4.25) and (4.28). □

4.2 Existence and regularity for the first eigenfunction of the higher order fractional Laplacian

The goal of these pages is to study the boundary behavior of the first Dirichlet eigenfunction of higher order fractional equations.

For this, writing $s = m + \sigma$, with $m \in \mathbb{N}$ and $\sigma \in (0, 1)$, we define the energy space

$$H^s_0(B_1) := \{u \in H^s(\mathbb{R}^n);\ u = 0 \text{ in } \mathbb{R}^n \setminus B_1\}, \tag{4.29}$$

endowed with the Hilbert norm

$$\|u\|_{H^s_0(B_1)} := \left(\sum_{|\alpha| \leq m} \|\partial^\alpha u\|^2_{L^2(B_1)} + \mathcal{E}_s(u, u) \right)^{\frac{1}{2}}, \tag{4.30}$$

where

$$\mathcal{E}_s(u, v) = \int_{\mathbb{R}^n} |\xi|^{2s} \mathcal{F}u(\xi) \overline{\mathcal{F}v(\xi)}\, d\xi, \tag{4.31}$$

with $\mathcal{F}$ being the Fourier transform and using the notation $\overline{z}$ to denote the complex conjugate of a complex number z.

In this setting, we consider $u \in H^s_0(B_1)$ to be such that

$$\begin{cases} (-\Delta)^s u = \lambda_1 u & \text{in } B_1, \\ u = 0 & \text{in } \mathbb{R}^n \setminus \overline{B_1}, \end{cases} \tag{4.32}$$

for every $s > 0$, with λ_1 as small as possible.

The existence of solutions of (4.32) is ensured via variational techniques, as stated in the following result.

Lemma 4.3. *The functional $\mathcal{E}_s(u, u)$ attains its minimum λ_1 on the functions in $H^s_0(B_1)$ with unit norm in $L^2(B_1)$.*

The minimizer satisfies (4.32).
In addition, $\lambda_1 > 0$.

Proof. The proof is based on the direct method in the calculus of variations. We provide some details for completeness. Let $s = m + \sigma$, with $m \in \mathbb{N}$ and $\sigma \in (0, 1)$. Let us consider a minimizing sequence $u_j \in H_0^s(B_1) \subseteq H^m(\mathbb{R}^n)$ such that $\|u_j\|_{L^2(B_1)} = 1$ and

$$\lim_{j\to+\infty} \mathcal{E}_s(u_j, u_j) = \inf_{\substack{u\in H_0^s(B_1)\\ \|u\|_{L^2(B_1)}=1}} \mathcal{E}_s(u, u).$$

In particular, u_j is bounded in $H_0^s(B_1)$ uniformly in j, so, up to a subsequence, it converges to some $u_\star$ weakly in $H_0^s(B_1)$ and strongly in $L^2(B_1)$ as $j \to +\infty$.

The weak lower semicontinuity of the seminorm $\mathcal{E}_s(\cdot, \cdot)$ then implies that $u_\star$ is the desired minimizer.

Then, given $\phi \in C_0^\infty(B_1)$, we have

$$\mathcal{E}_s(u_\star + \epsilon\phi, u_\star + \epsilon\phi) \geq \mathcal{E}_s(u_\star, u_\star),$$

for every $\epsilon \in \mathbb{R}$, and therefore (4.32) is satisfied in the sense of distributions, and also in the classical sense by the elliptic regularity theory.

Finally, we have $\mathcal{E}_s(u_\star, u_\star) > 0$, since $u_\star$ (and thus $\mathcal{F}u_\star$) does not vanish identically. Consequently,

$$\lambda_1 = \frac{\mathcal{E}_s(u_\star, u_\star)}{\|u_\star\|_{L^2(B_1)}^2} = \mathcal{E}_s(u_\star, u_\star) > 0,$$

as desired. □

Our goal is now to apply Proposition 4.2 to solutions of (4.32), taking $f := \lambda u$. To this end, we have to check that condition (4.18) is satisfied, namely, that solutions of (4.32) are Hölder continuous in $B_1 \setminus B_r$, for any $0 < r < 1$.

To this aim, we prove that polyharmonic operators of any order $s > 0$ always admit a first eigenfunction in the ball which does not change sign and which is radially symmetric. For this, we start by discussing the sign property.

Lemma 4.4. *There exists a nontrivial solution of* (4.32) *that does not change sign.*

Proof. We exploit a method explained in detail in Section 3.1 of [32]. As a matter of fact, when $s \in \mathbb{N}$, the desired result is exactly Theorem 3.7 in [32].

Let u be as in Lemma 4.3. If either $u \geq 0$ or $u \leq 0$, then the desired result is proved. Hence, we argue by contradiction, assuming that u attains strictly positive and strictly negative values. We define

$$\mathcal{K} := \{w : \mathbb{R}^n \to \mathbb{R} \text{ s.t. } \mathcal{E}_s(w, w) < +\infty, \text{ and } w \geq 0 \text{ in } B_1\}.$$

Also, we set

$$\mathcal{K}^\star := \{w \in H_0^s(B_1) \text{ s. t. } \mathcal{E}_s(w,v) \le 0 \text{ for all } v \in \mathcal{K}\}.$$

We claim that

$$\text{if } w \in \mathcal{K}^\star, \text{ then } w \le 0. \tag{4.33}$$

To prove this, we recall the notation in (4.1), take $\phi \in C_0^\infty(B_1) \cap \mathcal{K}$, and let

$$v_\phi(x) := \begin{cases} \int_{B_1} \mathcal{G}_s(x,y)\phi(y)\,dy & \text{if } x \in B_1, \\ 0 & \text{if } x \in \mathbb{R}^n \setminus B_1. \end{cases}$$

Then $v_\phi \in \mathcal{K}$ and it satisfies $(-\Delta)^s v_\phi = \phi$ in B_1, thanks to [24] or [2].

Consequently, we can write, for every $x \in B_1$,

$$\phi(x) = \mathcal{F}^{-1}(|\xi|^{2s}\mathcal{F}v_\phi)(x).$$

Hence, for every $w \in \mathcal{K}^\star$,

$$\begin{aligned} 0 &\ge \mathcal{E}_s(w, v_\phi) \\ &= \int_{\mathbb{R}^n} |\xi|^{2s}\mathcal{F}v_\phi(\xi)\overline{\mathcal{F}w(\xi)}\,d\xi \\ &= \int_{\mathbb{R}^n} \mathcal{F}^{-1}(|\xi|^{2s}\mathcal{F}v_\phi)(x)w(x)\,dx \\ &= \int_{B_1} \mathcal{F}^{-1}(|\xi|^{2s}\mathcal{F}v_\phi)(x)w(x)\,dx \\ &= \int_{B_1} \phi(x)w(x)\,dx. \end{aligned}$$

Since ϕ is arbitrary and nonnegative, this gives $w \le 0$, and this establishes (4.33).

Furthermore, by Theorem 3.4 in [32], we write

$$u = u_1 + u_2,$$

with $u_1 \in \mathcal{K} \setminus \{0\}$, $u_2 \in \mathcal{K}^\star \setminus \{0\}$, and $\mathcal{E}_s(u_1, u_2) = 0$.

We observe that

$$\mathcal{E}_s(u_1 - u_2, u_1 - u_2) = \mathcal{E}_s(u_1, u_1) + \mathcal{E}_s(u_2, u_2) + 2\mathcal{E}_s(u_1, u_2) = \mathcal{E}_s(u_1, u_1) + \mathcal{E}_s(u_2, u_2).$$

In the same way,

$$\mathcal{E}_s(u, u) = \mathcal{E}_s(u_1 + u_2, u_1 + u_2) = \mathcal{E}_s(u_1, u_1) + \mathcal{E}_s(u_2, u_2),$$

and therefore

$$\mathcal{E}_s(u_1 - u_2, u_1 - u_2) = \mathcal{E}_s(u, u). \tag{4.34}$$

However,

$$\begin{aligned}\|u_1 - u_2\|^2_{L^2(B_1)} - \|u\|^2_{L^2(B_1)} &= \|u_1 - u_2\|^2_{L^2(B_1)} - \|u_1 + u_2\|^2_{L^2(B_1)} \\ &= -4 \int_{B_1} u_1(x)u_2(x)\,dx.\end{aligned}$$

As a consequence, since $u_2 \le 0$ in view of (4.33), we conclude that

$$\|u_1 - u_2\|^2_{L^2(B_1)} - \|u\|^2_{L^2(B_1)} \ge 0.$$

From this and (4.34) it follows that the function $u_1 - u_2$ is also a minimizer for the variational problem in Lemma 4.3. Since now $u_1 - u_2 \ge 0$, the desired result follows. □

Now, we define the spherical mean of a function v by

$$v_\sharp(x) := \frac{1}{|\mathbb{S}^{n-1}|} \int_{\mathbb{S}^{n-1}} v(\mathcal{R}_\omega x)\,d\mathcal{H}^{n-1}(\omega),$$

where $\mathcal{R}_\omega$ is the rotation corresponding to the solid angle $\omega \in \mathbb{S}^{n-1}$, $\mathcal{H}^{n-1}$ is the standard Hausdorff measure, and $|\mathbb{S}^{n-1}| = \mathcal{H}^{n-1}(\mathbb{S}^{n-1})$. Note that $v_\sharp(x) = v_\sharp(\mathcal{R}_\varpi x)$ for any $\varpi \in \mathbb{S}^{n-1}$, that is, $v_\sharp$ is rotationally invariant.

Then, we have the following.

Lemma 4.5. *Any positive power of the Laplacian commutes with the spherical mean, that is,*

$$((-\Delta)^s v)_\sharp(x) = (-\Delta)^s v_\sharp(x).$$

Proof. By density, we prove the claim for a function v in the Schwartz space of smooth and rapidly decreasing functions. In this setting, writing $\mathcal{R}^T_\omega$ to denote the transpose of the rotation $\mathcal{R}_\omega$ and changing variable $\eta := \mathcal{R}^T_\omega \xi$, we have

$$\begin{aligned}(-\Delta)^s v(\mathcal{R}_\omega x) &= \int_{\mathbb{R}^n} |\xi|^{2s} \mathcal{F}v(\xi) e^{2\pi i \mathcal{R}_\omega x\cdot\xi}\,d\xi \\ &= \int_{\mathbb{R}^n} |\xi|^{2s} \mathcal{F}v(\xi) e^{2\pi i x\cdot \mathcal{R}^T_\omega \xi}\,d\xi \\ &= \int_{\mathbb{R}^n} |\eta|^{2s} \mathcal{F}v(\mathcal{R}_\omega \eta) e^{2\pi i x\cdot\eta}\,d\eta.\end{aligned} \tag{4.35}$$

However, using the substitution $y := \mathcal{R}^T_\omega x$,

$$\begin{aligned}\mathcal{F}v(\mathcal{R}_\omega\eta) &= \int_{\mathbb{R}^n} v(x)e^{-2\pi i x\cdot\mathcal{R}_\omega\eta}\,dx\\ &= \int_{\mathbb{R}^n} v(x)e^{-2\pi i\mathcal{R}_\omega^T x\cdot\eta}\,dx\\ &= \int_{\mathbb{R}^n} v(\mathcal{R}_\omega y)e^{-2\pi i y\cdot\eta}\,dy,\end{aligned}$$

and therefore, recalling (4.35),

$$(-\Delta)^s v(\mathcal{R}_\omega x) = \iint_{\mathbb{R}^n\times\mathbb{R}^n} |\eta|^{2s} v(\mathcal{R}_\omega y)e^{2\pi i(x-y)\cdot\eta}\,dy\,d\eta.$$

As a consequence,

$$\begin{aligned}((-\Delta)^s v)_\sharp(x) &= \frac{1}{|\mathbb{S}^{n-1}|}\int_{\mathbb{S}^{n-1}} (-\Delta)^s v(\mathcal{R}_\omega x)\,d\mathcal{H}^{n-1}(\omega)\\ &= \frac{1}{|\mathbb{S}^{n-1}|}\iiint_{\mathbb{S}^{n-1}\times\mathbb{R}^n\times\mathbb{R}^n} |\eta|^{2s} v(\mathcal{R}_\omega y)e^{2\pi i(x-y)\cdot\eta}\,d\mathcal{H}^{n-1}(\omega)\,dy\,d\eta\\ &= \iint_{\mathbb{R}^n\times\mathbb{R}^n} |\eta|^{2s} v_\sharp(y)e^{2\pi i(x-y)\cdot\eta}\,dy\,d\eta\\ &= \int_{\mathbb{R}^n} |\eta|^{2s}\mathcal{F}(v_\sharp)(\eta)e^{2\pi i x\cdot\eta}\,d\eta\\ &= (-\Delta)^s v_\sharp(x),\end{aligned}$$

as desired. □

It is also useful to observe that the spherical mean is compatible with the energy bounds. In particular, we have the following observation.

Lemma 4.6. *We have*

$$\mathcal{E}_s(v_\sharp, v_\sharp) \le \mathcal{E}_s(v, v). \tag{4.36}$$

Moreover,

$$\text{if } v \in H_0^s(B_1), \quad \text{then so does } v_\sharp. \tag{4.37}$$

Proof. We see that

$$\begin{aligned}\mathcal{F}(v_\sharp)(\xi) &= \int_{\mathbb{R}^n} v_\sharp(x)e^{-2\pi i x\cdot\xi}\,dx\\ &= \frac{1}{|\mathbb{S}^{n-1}|}\iint_{\mathbb{S}^{n-1}\times\mathbb{R}^n} v(\mathcal{R}_\omega x)e^{-2\pi i x\cdot\xi}\,d\mathcal{H}^{n-1}(\omega)\,dx\end{aligned}$$

and therefore, taking the complex conjugated,

$$\overline{\mathcal{F}(v_\sharp)(\xi)} = \frac{1}{|\mathbb{S}^{n-1}|} \iint_{\mathbb{S}^{n-1}\times\mathbb{R}^n} v(\mathcal{R}_\omega x) e^{2\pi i x\cdot\xi}\, d\mathcal{H}^{n-1}(\omega)\, dx.$$

Hence, by (4.31), and exploiting the changes of variables $y := \mathcal{R}_\omega x$ and $\tilde{y} := \mathcal{R}_{\tilde{\omega}}\tilde{x}$,

$$\begin{aligned}
&\mathcal{E}_s(v_\sharp, v_\sharp)\\
&= \int_{\mathbb{R}^n} |\xi|^{2s} \mathcal{F}(v_\sharp)(\xi)\overline{\mathcal{F}(v_\sharp)(\xi)}\, d\xi\\
&= \frac{1}{|\mathbb{S}^{n-1}|^2} \iiiint\int_{\mathbb{S}^{n-1}\times\mathbb{S}^{n-1}\times\mathbb{R}^n\times\mathbb{R}^n\times\mathbb{R}^n} |\xi|^{2s} v(\mathcal{R}_\omega x) v(\mathcal{R}_{\tilde{\omega}}\tilde{x})\\
&\quad\times e^{2\pi i(\tilde{x}-x)\cdot\xi}\, d\mathcal{H}^{n-1}(\omega)\, d\mathcal{H}^{n-1}(\tilde{\omega})\, dx\, d\tilde{x}\, d\xi\\
&= \frac{1}{|\mathbb{S}^{n-1}|^2} \iiiint\int_{\mathbb{S}^{n-1}\times\mathbb{S}^{n-1}\times\mathbb{R}^n\times\mathbb{R}^n\times\mathbb{R}^n} |\xi|^{2s} v(y) v(\tilde{y})\\
&\quad\times e^{2\pi i\tilde{y}\cdot\mathcal{R}_{\tilde{\omega}}\xi} e^{-2\pi i y\cdot\mathcal{R}_\omega\xi}\, d\mathcal{H}^{n-1}(\omega)\, d\mathcal{H}^{n-1}(\tilde{\omega})\, dy\, d\tilde{y}\, d\xi\\
&= \frac{1}{|\mathbb{S}^{n-1}|^2} \iiint_{\mathbb{S}^{n-1}\times\mathbb{S}^{n-1}\times\mathbb{R}^n} |\xi|^{2s} \mathcal{F}v(\mathcal{R}_\omega\xi)\overline{\mathcal{F}v(\mathcal{R}_{\tilde{\omega}}\xi)}\, d\mathcal{H}^{n-1}(\omega)\, d\mathcal{H}^{n-1}(\tilde{\omega})\, d\xi.
\end{aligned}$$

Consequently, using the Cauchy–Schwarz inequality and the substitutions $\eta := \mathcal{R}_\omega\xi$ and $\tilde{\eta} := \mathcal{R}_{\tilde{\omega}}\xi$,

$$\begin{aligned}
\mathcal{E}_s(v_\sharp, v_\sharp) &\le \frac{1}{|\mathbb{S}^{n-1}|^2} \iiint_{\mathbb{S}^{n-1}\times\mathbb{S}^{n-1}\times\mathbb{R}^n} |\xi|^{2s} |\mathcal{F}v(\mathcal{R}_\omega\xi)||\mathcal{F}v(\mathcal{R}_{\tilde{\omega}}\xi)|\, d\mathcal{H}^{n-1}(\omega)\, d\mathcal{H}^{n-1}(\tilde{\omega})\, d\xi\\
&\le \frac{1}{|\mathbb{S}^{n-1}|^2}\Big(\iiint_{\mathbb{S}^{n-1}\times\mathbb{S}^{n-1}\times\mathbb{R}^n} |\xi|^{2s} |\mathcal{F}v(\mathcal{R}_\omega\xi)|^2\, d\mathcal{H}^{n-1}(\omega)\, d\mathcal{H}^{n-1}(\tilde{\omega})\, d\xi\Big)^{\frac{1}{2}}\\
&\quad\cdot\Big(\iiint_{\mathbb{S}^{n-1}\times\mathbb{S}^{n-1}\times\mathbb{R}^n} |\xi|^{2s} |\mathcal{F}v(\mathcal{R}_{\tilde{\omega}}\xi)|^2\, d\mathcal{H}^{n-1}(\omega)\, d\mathcal{H}^{n-1}(\tilde{\omega})\, d\xi\Big)^{\frac{1}{2}}\\
&= \frac{1}{|\mathbb{S}^{n-1}|^2}\Big(\iiint_{\mathbb{S}^{n-1}\times\mathbb{S}^{n-1}\times\mathbb{R}^n} |\eta|^{2s} |\mathcal{F}v(\eta)|^2\, d\mathcal{H}^{n-1}(\omega)\, d\mathcal{H}^{n-1}(\tilde{\omega})\, d\eta\Big)^{\frac{1}{2}}\\
&\quad\cdot\Big(\iiint_{\mathbb{S}^{n-1}\times\mathbb{S}^{n-1}\times\mathbb{R}^n} |\tilde{\eta}|^{2s} |\mathcal{F}v(\tilde{\eta})|^2\, d\mathcal{H}^{n-1}(\omega)\, d\mathcal{H}^{n-1}(\tilde{\omega})\, d\tilde{\eta}\Big)^{\frac{1}{2}}\\
&= \Big(\int_{\mathbb{R}^n} |\eta|^{2s} |\mathcal{F}v(\eta)|^2\, d\eta\Big)^{\frac{1}{2}} \Big(\int_{\mathbb{R}^n} |\tilde{\eta}|^{2s} |\mathcal{F}v(\tilde{\eta})|^2\, d\tilde{\eta}\Big)^{\frac{1}{2}}\\
&= \mathcal{E}_s(v, v).
\end{aligned}$$

This proves (4.36).

Now, we prove (4.37). For this, we observe that

$$\frac{\partial^\ell v_\sharp}{\partial x_{j_1}\cdots\partial x_{j_\ell}}(x) = \frac{1}{|\mathbb{S}^{n-1}|}\sum_{k_1,\dots,k_\ell=1}^{n}\int_{\mathbb{S}^{n-1}} \frac{\partial^\ell v}{\partial x_{k_1}\cdots\partial x_{k_\ell}}(\mathcal{R}_\omega x)\mathcal{R}_\omega^{k_1 j_1}\cdots\mathcal{R}_\omega^{k_\ell j_\ell}\, d\mathcal{H}^{n-1}(\omega),$$

for every $\ell \in \mathbb{N}$ and $j_1,\dots,j_\ell \in \{1,\dots,n\}$, where $\mathcal{R}_\omega^{jk}$ denotes the (j,k) component of the matrix $\mathcal{R}_\omega$. In particular,

$$\left|\frac{\partial^\ell v_\sharp}{\partial x_{j_1}\cdots\partial x_{j_\ell}}(x)\right| \le C \sum_{k_1,\dots,k_\ell=1}^{n}\int_{\mathbb{S}^{n-1}} \left|\frac{\partial^\ell v}{\partial x_{k_1}\cdots\partial x_{k_\ell}}(\mathcal{R}_\omega x)\right| d\mathcal{H}^{n-1}(\omega),$$

for some $C>0$ only depending on n and ℓ, and hence

$$\begin{aligned}\left\|\frac{\partial^\ell v_\sharp}{\partial x_{j_1}\cdots\partial x_{j_\ell}}(x)\right\|_{L^2(B_1)}^2 &\le C \sum_{k_1,\dots,k_\ell=1}^{n}\iint_{\mathbb{S}^{n-1}\times B_1} \left|\frac{\partial^\ell v}{\partial x_{k_1}\cdots\partial x_{k_\ell}}(\mathcal{R}_\omega x)\right|^2 d\mathcal{H}^{n-1}(\omega)\,dx\\ &= C \sum_{k_1,\dots,k_\ell=1}^{n}\iint_{\mathbb{S}^{n-1}\times B_1} \left|\frac{\partial^\ell v}{\partial x_{k_1}\cdots\partial x_{k_\ell}}(y)\right|^2 d\mathcal{H}^{n-1}(\omega)\,dy\\ &= C \sum_{k_1,\dots,k_\ell=1}^{n}\left\|\frac{\partial^\ell v}{\partial x_{k_1}\cdots\partial x_{k_\ell}}\right\|_{L^2(B_1)}^2,\end{aligned}$$

up to renaming C.

This, together with (4.30) and (4.36), gives (4.37), as desired. □

With this preliminary work, we can now find a nontrivial, nonnegative, and radial solution of (4.32).

Proposition 4.7. *There exists a solution of* (4.32) *in* $H_0^s(B_1)$ *which is radial, nonnegative, and with unit norm in* $L^2(B_1)$.

Proof. Let $u \ge 0$ be a nontrivial solution of (4.32) whose existence is warranted by Lemma 4.4.

Then $u_\sharp \ge 0$. Moreover,

$$\begin{aligned}\int_{B_1} u_\sharp(x)\,dx &= \frac{1}{|\mathbb{S}^{n-1}|}\iint_{\mathbb{S}^{n-1}\times B_1} u(\mathcal{R}_\omega x)\,d\mathcal{H}^{n-1}(\omega)\,dx\\ &= \frac{1}{|\mathbb{S}^{n-1}|}\iint_{\mathbb{S}^{n-1}\times B_1} u(y)\,d\mathcal{H}^{n-1}(\omega)\,dy = \int_{B_1} u(y)\,dy > 0,\end{aligned}$$

and therefore $u_\sharp$ does not vanish identically.

As a consequence, we can define

$$u_\star := \frac{u_\sharp}{\|u_\sharp\|_{L^2(B_1)}}.$$

We know that $u_\star \in H_0^s(B_1)$, due to (4.37). Moreover, in view of Lemma 4.5,

$$(-\Delta)^s u_\star = \frac{(-\Delta)^s u_\sharp}{\|u_\sharp\|_{L^2(B_1)}} = \frac{((-\Delta)^s u)_\sharp}{\|u_\sharp\|_{L^2(B_1)}} = \frac{\lambda_1 u_\sharp}{\|u_\sharp\|_{L^2(B_1)}} = \lambda_1 u_\star,$$

which gives the desired result. □

Now, we are in the position of proving the following result.

Lemma 4.8. *Let $s \geq 1$ and $r \in (0,1)$. If $u \in H_0^s(B_1)$ and u is radial, then $u \in C^\alpha(\mathbb{R}^n \setminus B_r)$ for any $\alpha \in [0, \frac{1}{2}]$.*

Proof. We write

$$u(x) = u_0(|x|), \quad \text{for some } u_0 : [0,+\infty) \to \mathbb{R} \tag{4.38}$$

and we observe that $u \in H_0^s(B_1) \subset H^1(\mathbb{R}^n)$.

Accordingly, for any $0 < r < 1$, we have

$$\infty > \int_{\mathbb{R}^n \setminus B_r} |u(x)|^2 \, dx = \int_r^{+\infty} |u_0(\rho)|^2 \rho^{n-1} \, d\rho \geq r^{n-1} \int_r^{+\infty} |u_0(\rho)|^2 \, d\rho \tag{4.39}$$

and

$$\infty > \int_{\mathbb{R}^n \setminus B_r} |\nabla u(x)|^2 \, dx = \int_r^{+\infty} |u_0'(\rho)|^2 \rho^{n-1} \, d\rho \geq r^{n-1} \int_r^{+\infty} |u_0'(\rho)|^2 \, d\rho. \tag{4.40}$$

Thanks to (4.39) and (4.40), we have $u_0 \in H^1((r,+\infty))$, with $u_0 = 0$ in $[1,+\infty)$.

Then, from the Morrey Embedding Theorem, it follows that $u_0 \in C^\alpha((r,+\infty))$ for any $\alpha \in [0, \frac{1}{2}]$, which leads to the desired result. □

Corollary 4.9. *Let $s \in (0,+\infty)$. There exists a radial, nonnegative, and nontrivial solution of (4.32) which belongs to $H_0^s(B_1) \cap C^\alpha(\mathbb{R}^n \setminus B_{1/2})$, for some $\alpha \in (0,1)$.*

Proof. If $s \in (0,1)$, the desired claim follows from Corollary 8 in [26].

If instead $s \geq 1$, we obtain the desired result as a consequence of Proposition 4.7 and Lemma 4.8. □

4.3 Boundary asymptotics of the first eigenfunctions of $(-\Delta)^s$

In Lemma 4 of [26], some precise asymptotics at the boundary for the first Dirichlet eigenfunction of $(-\Delta)^s$ have been established in the range $s \in (0,1)$.

Here, we obtain a related expansion in the range $s > 0$ for the eigenfunction provided in Corollary 4.9. The result that we obtain is the following.

Proposition 4.10. *There exists a nontrivial solution ϕ_* of (4.32) which belongs to $H_0^s(B_1) \cap C^\alpha(\mathbb{R}^n \setminus B_{1/2})$, for some $\alpha \in (0,1)$, such that, for every $e \in \partial B_1$ and $\beta = (\beta_1, \dots, \beta_n) \in \mathbb{N}^n$,*

$$\lim_{\epsilon \searrow 0} \epsilon^{|\beta|-s} \partial^\beta \phi_*(e + \epsilon X) = (-1)^{|\beta|} k_* s(s-1) \cdots (s - |\beta| + 1) e_1^{\beta_1} \cdots e_n^{\beta_n} (-e \cdot X)_+^{s-|\beta|},$$

in the sense of distribution, with $|\beta| := \beta_1 + \dots + \beta_n$ and $k_* > 0$.

The proof of Proposition 4.10 relies on Proposition 4.2 and some auxiliary computations on the Green function in (4.1). We start with the following result.

Lemma 4.11. *Let $0 < r < 1$, $e \in \partial B_1$, $s > 0$, $f \in C^\alpha(\mathbb{R}^n \setminus B_r) \cap L^2(\mathbb{R}^n)$ for some $\alpha \in (0,1)$, and $f = 0$ outside B_1. Then the integral*

$$\int_{B_1} f(z) \frac{(1 - |z|^2)^s}{s|z - e|^n} \, dz \tag{4.41}$$

is finite.

Proof. We denote by I the integral in (4.41). We let

$$I_1 := \int_{B_1 \setminus B_r} f(z) \frac{(1 - |z|^2)^s}{s|z - e|^n} \, dz \quad \text{and} \quad I_2 := \int_{B_r} f(z) \frac{(1 - |z|^2)^s}{s|z - e|^n} \, dz.$$

Then we have

$$I = I_1 + I_2. \tag{4.42}$$

Now, if $z \in B_1 \setminus B_r$, we have

$$f(z) \le |f(z) - f(e)| \le C|z - e|^\alpha,$$

and therefore

$$I_1 \le \int_{B_1 \setminus B_r} \frac{(1 - |z|^2)^s}{s|z - e|^{n-\alpha}} \, dz < \infty. \tag{4.43}$$

If instead $z \in B_r$,

$$|z - e| \ge 1 - r > 0,$$

and consequently

$$I_2 \le \frac{1}{s(1 - r)^n} \int_{B_r} f(z) \, dz < \infty. \tag{4.44}$$

The desired result follows from (4.42), (4.43), and (4.44). □

The next result gives a precise boundary behavior of the Green function for any $s > 0$ (the case in which $s \in (0,1)$ and $f \in C^{\alpha}(\mathbb{R}^n)$ was considered in Lemma 6 of [26], and in fact the proof presented here also simplifies the one in Lemma 6 of [26] for the setting considered there).

Lemma 4.12. *Let $e, \omega \in \partial B_1$, $\epsilon_0 > 0$, and $r \in (0,1)$. Assume that*

$$e + \epsilon\omega \in B_1, \tag{4.45}$$

for any $\epsilon \in (0,\epsilon_0]$. Let $f \in C^{\alpha}(\mathbb{R}^n \setminus B_r) \cap L^2(\mathbb{R}^n)$ for some $\alpha \in (0,1)$, with $f = 0$ outside B_1.

Then

$$\lim_{\epsilon\searrow 0} \epsilon^{-s} \int_{B_1} f(z)\mathcal{G}_s(e+\epsilon\omega, z)\,dz = k(n,s)\int_{B_1} f(z)\frac{(-2e\cdot\omega)^s(1-|z|^2)^s}{s|z-e|^n}\,dz, \tag{4.46}$$

for a suitable normalizing constant $k(n,s) > 0$.

Proof. In light of (4.45), we have

$$1 > |e+\epsilon\omega|^2 = 1 + \epsilon^2 + 2\epsilon e\cdot\omega,$$

and therefore

$$-e\cdot\omega > \frac{\epsilon}{2} > 0. \tag{4.47}$$

Moreover, if r_0 is as given in (4.1), we have, for all $z \in B_1$,

$$r_0(e+\epsilon\omega, z) = \frac{\epsilon(-\epsilon - 2e\cdot\omega)(1-|z|^2)}{|z-e-\epsilon\omega|^2} \le \frac{3\epsilon}{|z-e-\epsilon\omega|^2}. \tag{4.48}$$

Also, a Taylor series representation allows us to write, for any $t \in (-1,1)$,

$$\frac{t^{s-1}}{(t+1)^{\frac{n}{2}}} = \sum_{k=0}^{\infty}\binom{-n/2}{k}t^{k+s-1}. \tag{4.49}$$

We also note that

$$\begin{aligned}\left|\binom{-n/2}{k}\right| &= \left|\frac{-\frac{n}{2}(-\frac{n}{2}-1)\cdots(-\frac{n}{2}-k+1)}{k!}\right| = \frac{\frac{n}{2}(\frac{n}{2}+1)\cdots(\frac{n}{2}+(k-1))}{k!}\\ &\le \frac{n(n+1)\cdots(n+(k-1))}{k!} \le \frac{(n+(k-1))!}{k!} = (k+1)\cdots(n+(k-1))\\ &\le (n+k+1)^{n+1}.\end{aligned} \tag{4.50}$$

From this and the Root Test it follows that the series in (4.49) is uniformly convergent on compact sets in $(-1,1)$.

As a consequence, if we set

$$r_1(x,z) := \min\left\{\frac{1}{2}, r_0(x,z)\right\}, \tag{4.51}$$

we can switch integration and summation signs and obtain

$$\int_0^{r_1(x,z)} \frac{t^{s-1}}{(t+1)^{\frac{n}{2}}}\,dt = \sum_{k=0}^{\infty} c_k\big(r_1(x,z)\big)^{k+s}, \tag{4.52}$$

where

$$c_k := \frac{1}{k+s}\binom{-n/2}{k}.$$

Once again, from the bound in (4.50), together with (4.51), it follows that the series in (4.52) is convergent.

Now, we omit for simplicity the normalizing constant $k(n,s)$ in the definition of the Green function in (4.1), and we define

$$\mathcal{G}(x,z) := |z-x|^{2s-n} \sum_{k=0}^{\infty} c_k\big(r_1(x,z)\big)^{k+s} \tag{4.53}$$

and

$$g(x,z) := |z-x|^{2s-n} \int_{r_1(x,z)}^{r_0(x,z)} \frac{t^{s-1}}{(t+1)^{\frac{n}{2}}}\,dt.$$

Using (4.1) and (4.52), and dropping dimensional constants for the sake of shortness, we write

$$\mathcal{G}_s(x,z) = \mathcal{G}(x,z) + g(x,z). \tag{4.54}$$

Now, we show that

$$g(x,z) \le \begin{cases} C\chi(r,z)|z-x|^{2s-n} & \text{if } n > 2s,\\ C\chi(r,z)\log r_0(x,z) & \text{if } n = 2s,\\ C\chi(r,z)|z-x|^{2s-n}(r_0(x,z))^{s-\frac{n}{2}} & \text{if } n < 2s, \end{cases} \tag{4.55}$$

where $\chi(r,z) = 1$ if $r_0(x,z) > \frac{1}{2}$ and $\chi(r,z) = 0$ if $r_0(x,z) \le \frac{1}{2}$. To check this, we note that if $r_0(x,z) \le \frac{1}{2}$, we have $r_1(x,z) = r_0(x,z)$, due to (4.51), and therefore $g(x,z) = 0$.

However, if $r_0(x,z) > \frac{1}{2}$, we deduce from (4.51) that $r_1(x,z) = \frac{1}{2}$, and consequently

$$g(x,z) \le |z-x|^{2s-n} \int_{1/2}^{r_0(x,z)} t^{s-\frac{n}{2}-1}\,dt \le \begin{cases} C|z-x|^{2s-n} & \text{if } n > 2s,\\ C\log r_0(x,z) & \text{if } n = 2s,\\ C|z-x|^{2s-n}(r_0(x,z))^{s-\frac{n}{2}} & \text{if } n < 2s, \end{cases}$$

for some constant $C > 0$. This completes the proof of (4.55).

Now, we exploit the bound in (4.55) when $x = e+\epsilon\omega$. For this, we note that if $r_0(e+\epsilon\omega, z) > \frac{1}{2}$, recalling (4.48), we find

$$|z-e-\epsilon\omega|^2 \le 6\epsilon < 9\epsilon, \tag{4.56}$$

and therefore $z \in B_{3\sqrt{\epsilon}}(e+\epsilon\omega)$.

Hence, using (4.55),

$$\begin{aligned} &\left|\int_{B_1} f(z)g(e+\epsilon\omega, z)dz\right| \\ &\quad \le \int_{B_{3\sqrt{\epsilon}}(e+\epsilon\omega)} |f(z)||g(e+\epsilon\omega,z)|dz \\ &\quad \le \begin{cases} C\int_{B_{3\sqrt{\epsilon}}(e+\epsilon\omega)} |f(z)||z-e-\epsilon\omega|^{2s-n}dz & \text{if } n>2s, \\ C\int_{B_{3\sqrt{\epsilon}}(e+\epsilon\omega)} |f(z)|\log r_0(e+\epsilon\omega,z)dz & \text{if } n=2s, \\ C\int_{B_{3\sqrt{\epsilon}}(e+\epsilon\omega)} |f(z)||z-e-\epsilon\omega|^{2s-n}(r_0(e+\epsilon\omega,z))^{s-\frac{n}{2}}dz & \text{if } n<2s. \end{cases} \end{aligned} \tag{4.57}$$

Now, if $z \in B_{3\sqrt{\epsilon}}(e+\epsilon\omega)$, then

$$|z-e| \le |z-e-\epsilon\omega| + |\epsilon\omega| \le 3\sqrt{\epsilon}+\epsilon < 4\sqrt{\epsilon}. \tag{4.58}$$

Furthermore, for a given $r \in (0,1)$, we have $B_{3\sqrt{\epsilon}}(e+\epsilon\omega) \subseteq \mathbb{R}^n \setminus B_r$, provided that ϵ is sufficiently small.

Hence, if $z \in B_{3\sqrt{\epsilon}}(e+\epsilon\omega)$, we can exploit the regularity of f and deduce that

$$|f(z)| = |f(z)-f(e)| \le C|z-e|^{\alpha}.$$

This and (4.58) lead to

$$|f(z)| \le C\epsilon^{\frac{\alpha}{2}}, \tag{4.59}$$

for every $z \in B_{3\sqrt{\epsilon}}(e+\epsilon\omega)$.

Thanks to (4.48), (4.57), and (4.59), we have

$$\begin{aligned} \left|\int_{B_1} f(z)g(e+\epsilon\omega,z)dz\right| &\le \begin{cases} C\epsilon^{\frac{\alpha}{2}}\int_{B_{3\sqrt{\epsilon}}(e+\epsilon\omega)} |z-e-\epsilon\omega|^{2s-n}dz & \text{if } n>2s, \\ C\epsilon^{\frac{\alpha}{2}}\int_{B_{3\sqrt{\epsilon}}(e+\epsilon\omega)} \log\frac{3\epsilon}{|z-e-\epsilon\omega|^2}dz & \text{if } n=2s, \\ C\epsilon^{\frac{\alpha}{2}+s-\frac{n}{2}}\int_{B_{3\sqrt{\epsilon}}(e+\epsilon\omega)} dz & \text{if } n<2s \end{cases} \\ &\le C\epsilon^{\frac{\alpha}{2}+s}, \end{aligned}$$

up to renaming C.

This and (4.54) give

$$\int_{B_1} f(z)\mathcal{G}_s(e+\epsilon\omega,z)dz = \int_{B_1} f(z)\mathcal{G}(e+\epsilon\omega,z)dz + o(\epsilon^s). \tag{4.60}$$

Now, we consider the series in (4.53), and we split the contribution coming from the index $k = 0$ from the ones coming from the indices $k > 0$, namely, we write

$$\begin{aligned}
&\mathcal{G}(x,z) = \mathcal{G}_0(x,z) + \mathcal{G}_1(x,z),\\
\text{with}\quad &\mathcal{G}_0(x,z) := \frac{|z-x|^{2s-n}}{s}(r_1(x,z))^s\\
\text{and}\quad &\mathcal{G}_1(x,z) := |z-x|^{2s-n}\sum_{k=1}^{+\infty} c_k(r_1(x,z))^{k+s}.
\end{aligned} \tag{4.61}$$

Firstly, we consider the contribution given by the term $\mathcal{G}_1$. Thanks to (4.51) and (4.59), we have

$$\begin{aligned}
&\Big|\int_{B_1\cap B_{3\sqrt{\epsilon}}(e+\epsilon\omega)} f(z)\mathcal{G}_1(e+\epsilon\omega,z)dz\Big|\\
&\quad\le \int_{B_{3\sqrt{\epsilon}}(e+\epsilon\omega)} |f(z)|\mathcal{G}_1(e+\epsilon\omega,z)dz\\
&\quad\le C\epsilon^{\frac{\alpha}{2}}\int_{B_{3\sqrt{\epsilon}}(e+\epsilon\omega)} |z-e-\epsilon\omega|^{2s-n}\sum_{k=1}^{+\infty}|c_k|(r_1(e+\epsilon\omega,z))^{k+s}dz\\
&\quad\le C\epsilon^{\frac{\alpha}{2}}\int_{B_{3\sqrt{\epsilon}}(e+\epsilon\omega)} |z-e-\epsilon\omega|^{2s-n}\sum_{k=1}^{+\infty}|c_k|\Big(\frac{1}{2}\Big)^{k+s}dz\\
&\quad\le C\epsilon^{\frac{\alpha}{2}}\int_{B_{3\sqrt{\epsilon}}(e+\epsilon\omega)} |z-e-\epsilon\omega|^{2s-n}dz\\
&\quad\le C\epsilon^{\frac{\alpha}{2}+s},
\end{aligned} \tag{4.62}$$

up to renaming the constant C step by step.

However, for every $z \in \mathbb{R}^n$,

$$|z| = |e+\epsilon\omega+z-e-\epsilon\omega| \ge |e+\epsilon\omega| - |z-e-\epsilon\omega| \ge 1-\epsilon-|z-e-\epsilon\omega|.$$

Therefore, for every $z \in B_1\setminus(B_r\cup B_{3\sqrt{\epsilon}}(e+\epsilon\omega))$, we can take $e_* := \frac{z}{|z|}$ and obtain

$$\begin{aligned}
|f(z)| &= |f(z)-f(e_*)| \le C|z-e_*|^\alpha = C(1-|z|)^\alpha\\
&\le C(\epsilon+|z-e-\epsilon\omega|)^\alpha \le C|z-e-\epsilon\omega|^\alpha,
\end{aligned} \tag{4.63}$$

up to renaming $C > 0$.

Also, using (4.48), we see that, for any $k > 0$,

$$(r_0(e+\epsilon\omega,z))^{s+\frac{\alpha}{4}}\Big(\frac{1}{2}\Big)^{k-\frac{\alpha}{4}} \le \frac{C\epsilon^{s+\frac{\alpha}{4}}}{2^k|z-e-\epsilon\omega|^{2s+\frac{\alpha}{2}}}. \tag{4.64}$$

This, from (4.51) and (4.63) it follows that if $z \in B_1 \setminus (B_r \cup B_{3\sqrt{\epsilon}}(e+\epsilon\omega))$, then

$$\begin{aligned}|f(z)\mathcal{G}_1(e+\epsilon\omega,z)| &\le C|z-e-\epsilon\omega|^{\alpha+2s-n}\sum_{k=1}^{+\infty}|c_k|(r_1(e+\epsilon\omega,z))^{k+s}\\ &= C|z-e-\epsilon\omega|^{\alpha+2s-n}\sum_{k=1}^{+\infty}|c_k|(r_1(e+\epsilon\omega,z))^{s+\frac{\alpha}{4}}(r_1(e+\epsilon\omega,z))^{k-\frac{\alpha}{4}}\\ &\le C|z-e-\epsilon\omega|^{\alpha+2s-n}\sum_{k=1}^{+\infty}|c_k|(r_0(e+\epsilon\omega,z))^{s+\frac{\alpha}{4}}\left(\frac{1}{2}\right)^{k-\frac{\alpha}{4}}\\ &\le C\epsilon^{s+\frac{\alpha}{4}}|z-e-\epsilon\omega|^{\frac{\alpha}{2}-n}\sum_{k=1}^{+\infty}\frac{|c_k|}{2^k},\end{aligned}$$

where the latter series is absolutely convergent thanks to (4.50).

This implies that, if we set $E := B_1 \setminus (B_r \cup B_{3\sqrt{\epsilon}}(e+\epsilon\omega))$, we have

$$\begin{aligned}\left|\int_E f(z)\mathcal{G}_1(e+\epsilon\omega,z)dz\right| &\le C\epsilon^{s+\frac{\alpha}{4}}\int_E |z-e-\epsilon\omega|^{\frac{\alpha}{2}-n}dz\\ &\le C\epsilon^{s+\frac{\alpha}{4}}\int_{B_1}|z-e-\epsilon\omega|^{\frac{\alpha}{2}-n}dz \le C\epsilon^{s+\frac{\alpha}{4}}\int_{B_3}|z|^{\frac{\alpha}{2}-n}dz \le C\epsilon^{s+\frac{\alpha}{4}}. \qquad (4.65)\end{aligned}$$

Moreover, if $z \in B_r$, we have

$$|e+\epsilon\omega-z| \ge 1-\epsilon-r,$$

and therefore, recalling (4.64),

$$\begin{aligned}\sup_{z\in B_r}|\mathcal{G}_1(e+\epsilon\omega,z)| &\le |z-e-\epsilon\omega|^{2s-n}\sum_{k=1}^{+\infty}|c_k|(r_1(e+\epsilon\omega,z))^{s+\frac{\alpha}{4}}(r_1(e+\epsilon\omega,z))^{k-\frac{\alpha}{4}}\\ &\le |z-e-\epsilon\omega|^{2s-n}\sum_{k=1}^{+\infty}|c_k|(r_0(e+\epsilon\omega,z))^{s+\frac{\alpha}{4}}\left(\frac{1}{2}\right)^{k-\frac{\alpha}{4}}\\ &\le C|z-e-\epsilon\omega|^{-n-\frac{\alpha}{2}}\sum_{k=1}^{+\infty}\frac{|c_k|}{2^k}\\ &\le C(1-\epsilon-r)^{-n-\frac{\alpha}{2}}\epsilon^{s+\frac{\alpha}{4}},\end{aligned}$$

up to renaming C.

As a consequence, we find

$$\begin{aligned}\left|\int_{B_r} f(z)\mathcal{G}_1(e+\epsilon\omega,z)dz\right| &\le \sup_{z\in B_r}|\mathcal{G}_1(e+\epsilon\omega,z)|\|f\|_{L^1(B_r)}\\ &\le \|f\|_{L^1(B_r)}(1-\epsilon-r)^{-n-\frac{\alpha}{2}}\epsilon^{s+\frac{\alpha}{4}}\\ &\le \|f\|_{L^1(B_r)}2^{n+\frac{\alpha}{2}}(1-r)^{-n-\frac{\alpha}{2}}\epsilon^{s+\frac{\alpha}{4}}\\ &= C\epsilon^{s+\frac{\alpha}{4}}, \qquad (4.66)\end{aligned}$$

as long as ϵ is suitably small with respect to r; C is a positive constant which depends on $\|f\|_{L^1(B_r)}$, r, n, and α.

Then, by (4.62), (4.65), and (4.66) we conclude that

$$\int_{B_1} f(z)\mathcal{G}_1(e+\epsilon\omega,z)dz = o(\epsilon^s). \tag{4.67}$$

Inserting this information into (4.60), and recalling (4.61), we obtain

$$\int_{B_1} f(z)\mathcal{G}_s(e+\epsilon\omega,z)dz = \int_{B_1} f(z)\mathcal{G}_0(e+\epsilon\omega,z)dz + o(\epsilon^s). \tag{4.68}$$

Now, we define

$$\mathcal{D}_1 := \{z \in B_1 \text{ s. t. } r_0(e+\epsilon\omega,z) > 1/2\}$$

and

$$\mathcal{D}_2 := \{z \in B_1 \text{ s. t. } r_0(e+\epsilon\omega,z) \le 1/2\}.$$

If $z \in \mathcal{D}_1$, then $z \in B_1 \setminus B_r$, thanks to (4.56), and hence we can use (4.57) and (4.59) and write

$$|f(z)\mathcal{G}_0(e+\epsilon\omega,z)| \le C\epsilon^{\frac{\alpha}{2}}|z-e-\epsilon\omega|^{2s-n}.$$

Then, recalling again (4.57),

$$\left|\int_{\mathcal{D}_1} f(z)\mathcal{G}_1(e+\epsilon\omega,z)dz\right| \le C\epsilon^{\frac{\alpha}{2}} \int_{B_{3\sqrt{\epsilon}}(e+\epsilon\omega)} |z-e-\epsilon\omega|^{2s-n}dz = C\epsilon^{\frac{\alpha}{2}+s}, \tag{4.69}$$

up to renaming the constant $C > 0$. This information and (4.68) give

$$\int_{B_1} f(z)\mathcal{G}_s(e+\epsilon\omega,z)dz = \int_{\mathcal{D}_2} f(z)\mathcal{G}_0(e+\epsilon\omega,z)dz + o(\epsilon^s).$$

Now, by (4.48) and (4.51), if $z \in \mathcal{D}_2$,

$$\mathcal{G}_0(e+\epsilon\omega,z) = \frac{|z-e-\epsilon\omega|^{2s-n}}{s}\big(r_0(e+\epsilon\omega)\big)^s = \frac{\epsilon^s(-\epsilon-2e\cdot\omega)^s(1-|z|^2)^s}{s|z-e-\epsilon\omega|^n}.$$

Hence, we have

$$\begin{aligned}
&\lim_{\epsilon\searrow 0}\epsilon^{-s}\int_{B_1} f(z)\mathcal{G}_s(e+\epsilon\omega,z)dz\\
&= \lim_{\epsilon\searrow 0}\epsilon^{-s}\int_{\mathcal{D}_2} f(z)\mathcal{G}_0(e+\epsilon\omega,z)dz\\
&= \lim_{\epsilon\searrow 0}\int_{\{2\epsilon(-\epsilon-2e\cdot\omega)(1-|z|^2)\le|z-e-\epsilon\omega|^2\}} f(z)\frac{(-\epsilon-2e\cdot\omega)^s(1-|z|^2)^s}{s|z-e-\epsilon\omega|^n}dz.
\end{aligned} \tag{4.70}$$

Now we set

$$F_\epsilon(z) := \begin{cases} f(z)\frac{(-\epsilon-2e\cdot\omega)^s(1-|z|^2)^s}{s|z-e-\epsilon\omega|^n} & \text{if } 2\epsilon(-\epsilon-2e\cdot\omega)(1-|z|^2) \le |z-e-\epsilon\omega|^2, \\ 0 & \text{otherwise,} \end{cases} \tag{4.71}$$

and we prove that for any $\eta > 0$ there exists $\delta > 0$ independent of ϵ such that, for any $E \subset \mathbb{R}^n$ with $|E| \le \delta$, we have

$$\int_{B_1\cap E} |F_\epsilon(z)|dz \le \eta. \tag{4.72}$$

To this aim, given η and E as above, we define

$$\rho := \min\left\{\epsilon(-\epsilon-2e\cdot\omega), \sqrt{2\epsilon(-\epsilon-2e\cdot\omega)(1-r)}, \left(\frac{2^{s+\alpha}s^2\epsilon^{s+\alpha}(-\epsilon-2e\cdot\omega)^\alpha\eta}{3^{2s}\|f\|_{C^\alpha(B_1\setminus B_r)}|\partial B_1|}\right)^{\frac{1}{2\alpha}}\right\}. \tag{4.73}$$

We stress that the above definition is well-posed, thanks to (4.47). In addition, using the integrability of f, we take $\delta > 0$ such that if $A \subseteq B_1$ and $|A| \le \delta$, then

$$\int_A |f(x)|\,dx \le \frac{s\rho^n\eta}{2\cdot 3^s}. \tag{4.74}$$

We set

$$E_1 := E \cap B_\rho(e+\epsilon\omega) \quad \text{and} \quad E_2 := E \setminus B_\rho(e+\epsilon\omega). \tag{4.75}$$

From (4.71), we see that

$$|F_\epsilon(z)| \le \frac{|f(z)|\chi_\star(z)}{2^s s\epsilon^s|z-e-\epsilon\omega|^{n-2s}},$$

where

$$\chi_\star(z) := \begin{cases} 1 & \text{if } 2\epsilon(-\epsilon-2e\cdot\omega)(1-|z|^2) \le |z-e-\epsilon\omega|^2, \\ 0 & \text{otherwise,} \end{cases}$$

and therefore

$$\int_{B_1\cap E_1} |F_\epsilon(z)|\,dz \le \int_{B_1\cap E_1} \frac{|f(z)|\chi_\star(z)}{2^s s\epsilon^s|z-e-\epsilon\omega|^{n-2s}}\,dz. \tag{4.76}$$

Now, for every $z \in B_1 \cap E_1 \subseteq B_\rho(e+\epsilon\omega)$ for which $\chi_\star(z) \ne 0$, we have

$$2\epsilon(-\epsilon-2e\cdot\omega)(1-|z|^2) \le |z-e-\epsilon\omega|^2 \le \rho^2,$$

and hence

$$|z| \geq \sqrt{1 - \frac{\rho^2}{2\epsilon(-\epsilon - 2e \cdot \omega)}} \geq 1 - \frac{\rho^2}{2\epsilon(-\epsilon - 2e \cdot \omega)},$$

which in turn gives $|z| \geq r$, recalling (4.73).

From this and (4.76) we deduce that

$$\begin{aligned}
\int_{B_1 \cap E_1} |F_\epsilon(z)|\, dz &\leq \int_{1 - \frac{\rho^2}{2\epsilon(-\epsilon-2e\cdot\omega)} \leq |z| < 1} \frac{\|f\|_{C^\alpha(B_1 \setminus B_r)} (1 - |z|)^\alpha}{2^s s \epsilon^s |z - e - \epsilon\omega|^{n-2s}}\, dz \\
&\leq \frac{\|f\|_{C^\alpha(B_1 \setminus B_r)}}{2^s s \epsilon^s} \left(\frac{\rho^2}{2\epsilon(-\epsilon - 2e \cdot \omega)} \right)^\alpha \int_{1 - \frac{\rho^2}{2\epsilon(-\epsilon-2e\cdot\omega)} \leq |z| < 1} \frac{dz}{|z - e - \epsilon\omega|^{n-2s}} \\
&\leq \frac{\|f\|_{C^\alpha(B_1 \setminus B_r)}}{2^s s \epsilon^s} \left(\frac{\rho^2}{2\epsilon(-\epsilon - 2e \cdot \omega)} \right)^\alpha \int_{B_3} \frac{dx}{|x|^{n-2s}} \\
&= \frac{3^{2s} \|f\|_{C^\alpha(B_1 \setminus B_r)} |\partial B_1|}{2^{s+\alpha+1} s^2 \epsilon^{s+\alpha} (-\epsilon - 2e \cdot \omega)^\alpha}\, \rho^{2\alpha} \\
&\leq \frac{\eta}{2},
\end{aligned} \tag{4.77}$$

where (4.73) has been exploited in the last inequality.

We also point out that, by (4.71), (4.74), and (4.75),

$$\begin{aligned}
\int_{B_1 \cap E_2} |F_\epsilon(z)|\, dz &\leq \int_{(B_1 \setminus B_\rho(e+\epsilon\omega)) \cap E} |f(z)| \frac{(-\epsilon - 2e \cdot \omega)^s (1 - |z|^2)^s}{s|z - e - \epsilon\omega|^n}\, dz \\
&\leq \frac{3^s}{s\rho^n} \int_{B_1 \cap E} |f(z)|\, dz \\
&\leq \frac{\eta}{2}.
\end{aligned}$$

This, (4.75), and (4.77) give (4.72), as desired.

Note also that the sequence $F_\epsilon(z)$ converges pointwise to the function

$$F(z) := f(z) \frac{(-2e \cdot \omega)^s (1 - |z|^2)^s}{s|z - e|^n}.$$

Hence (4.70), (4.72), and the Vitali Convergence Theorem allow us to conclude that

$$\begin{aligned}
\lim_{\epsilon \searrow 0} \int_{B_1} f(z) \mathcal{G}_s(e + \epsilon\omega, z) dz &= \lim_{\epsilon \searrow 0} \int_{B_1} F_\epsilon(z) dz \\
&= \int_{B_1} f(z) \frac{(-2e \cdot \omega)^s (1 - |z|^2)^s}{s|z - e|^n} dz,
\end{aligned} \tag{4.78}$$

which establishes the claim of Lemma 4.12 (note that the finiteness of the latter quantity in (4.78) follows from (4.11)). □

With this preliminary work, we can now establish the boundary behavior of solutions, which is needed in our setting. As a matter of fact, from Lemma 4.12 we immediately deduce the following.

Corollary 4.13. *Let $e, \omega \in \partial B_1$, $\epsilon_0 > 0$, and $r \in (0,1)$.*

Assume that $e + \epsilon\omega \in B_1$, for any $\epsilon \in (0, \epsilon_0]$. Let $f \in C^\alpha(\mathbb{R}^n \setminus B_r) \cap L^2(\mathbb{R}^n)$ for some $\alpha \in (0,1)$, with $f = 0$ outside B_1.

Let u be as in (4.19). Then

$$\lim_{\epsilon \searrow 0} \epsilon^{-s} u(e + \epsilon\omega) = k(n,s)(-2e \cdot \omega)^s \int_{B_1} f(z) \frac{(1 - |z|^2)^s}{s|z - e|^n} \, dz,$$

where $k(n,s)$ denotes a positive normalizing constant.

Now we apply the previous results to detect the boundary growth of a suitable first eigenfunction. For our purposes, the statement that we need is the following.

Corollary 4.14. *There exists a nontrivial solution ϕ_* of (4.32) which belongs to $H_0^s(B_1) \cap C^\alpha(\mathbb{R}^n \setminus B_{1/2})$, for some $\alpha \in (0,1)$, such that, for every $e \in \partial B_1$,*

$$\lim_{\epsilon \searrow 0} \epsilon^{-s} \phi_*(e + \epsilon\omega) = k_*(-e \cdot \omega)_+^s, \tag{4.79}$$

for a suitable constant $k_* > 0$.

Furthermore, for every $R \in (r, 1)$, there exists $C_R > 0$ such that

$$\sup_{x \in B_1 \setminus B_R} d^{-s}(x) |\phi_*(x)| \le C_R. \tag{4.80}$$

Proof. Let $\alpha \in (0,1)$ and $\phi \in H_0^s(B_1) \cap C^\alpha(\mathbb{R}^n \setminus B_{1/2})$ be the nonnegative and nontrivial solution of (4.32), as given in Corollary 4.9.

In the spirit of (4.19), we define

$$\phi_*(x) := \begin{cases} \lambda_1 \int_{B_1} \mathcal{G}_s(x,y)\phi(y)\, dy & \text{if } x \in B_1, \\ 0 & \text{if } x \in \mathbb{R}^n \setminus B_1. \end{cases}$$

We stress that we can use Proposition 4.2 in this context, with $f := \lambda_1 \phi$, since condition (4.18) is satisfied in this case.

Then, from (4.20) and (4.22), we know that $\phi_* \in H_0^s(B_1)$ and, from (4.21),

$$(-\Delta)^s \phi_* = \lambda_1 \phi \quad \text{in } B_1.$$

In particular, we have $(-\Delta)^s(\phi - \phi_*) = 0$ in B_1, and $\phi - \phi_* \in H_0^s(B_1)$, from which it follows that $\phi - \phi_*$ vanishes identically. Hence, we can write $\phi = \phi_*$, and thus ϕ_* is a solution of (4.32).

Now, we check (4.79). For this, we distinguish two cases. If $e \cdot \omega \geq 0$, we have

$$|e + \epsilon\omega|^2 = 1 + 2\epsilon e \cdot \omega + \epsilon^2 > 1,$$

for all $\epsilon > 0$. Then, in this case $e + \epsilon\omega \in \mathbb{R}^n \setminus B_1$, and therefore $\phi_*(e + \epsilon\omega) = 0$. This gives, in this case,

$$\lim_{\epsilon \searrow 0} \epsilon^{-s}\phi_*(e + \epsilon\omega) = 0. \tag{4.81}$$

If instead $e \cdot \omega < 0$, we see that

$$|e + \epsilon\omega|^2 = 1 + 2\epsilon e \cdot \omega + \epsilon^2 < 1,$$

for all $\epsilon > 0$ sufficiently small. Hence, we can exploit Corollary 4.13 and find that

$$\lim_{\epsilon \searrow 0} \epsilon^{-s}\phi_*(e + \epsilon\omega) = \lambda_1 k(n,s)(-2e \cdot \omega)^s \int_{B_1} \phi(z)\frac{(1 - |z|^2)^s}{s|z - e|^n}\, dz, \tag{4.82}$$

with $k(n,s) > 0$. Then, we define

$$k_* := 2^s k(n,s) \int_{B_1} \phi(z)\frac{(1 - |z|^2)^s}{s|z - e|^n}\, dz.$$

We observe that k_* is positive by construction, with $k(n,s) > 0$. Also, in light of Lemma 4.11, we know that k_* is finite. Hence, from (4.81) and (4.82) we obtain (4.79), as desired.

It only remains to check (4.80). For this, we use (4.21), and we see that, for every $R \in (r,1)$,

$$\sup_{x \in B_1 \setminus B_R} d^{-s}(x)|\phi_*(x)| \leq C_R \lambda_1 \big(\|\phi\|_{L^1(B_1)} + \|\phi\|_{L^\infty(B_1 \setminus B_r)}\big),$$

and this gives (4.80) up to renaming C_R. □

Now, we can complete the proof of Proposition 4.10, by arguing as follows.

Proof of Proposition 4.10. Let ψ be a test function in $C_0^\infty(\mathbb{R}^n)$. Let also $R := \frac{r+1}{2} \in (r,1)$ and

$$g_\epsilon(X) := \epsilon^{-s}\phi_*(e + \epsilon X)\partial^\beta \psi(X).$$

We claim that

$$\sup_{X \in \mathbb{R}^n} |g_\epsilon(X)| \leq C, \tag{4.83}$$

for some $C > 0$ independent of ϵ. To prove this, we distinguish three cases. If $e + \epsilon X \in \mathbb{R}^n \setminus B_1$, we have $\phi_*(e+\epsilon X) = 0$ and thus $g_\epsilon(X) = 0$. If instead $e+\epsilon X \in B_R$, we observe that

$$R > |e + \epsilon X| \geq 1 - \epsilon |X|,$$

and therefore $|X| \geq \frac{1-R}{\epsilon}$. In particular, in this case X falls outside the support of ψ, as long as $\epsilon > 0$ is sufficiently small, and consequently $\partial^\beta \psi(X) = 0$ and $g_\epsilon(X) = 0$.

Hence, to complete the proof of (4.83), we are only left with the case in which $e + \epsilon X \in B_1 \setminus B_R$. In this situation, we make use of (4.80) and we find that

$$\begin{aligned} |\phi_*(e + \epsilon X)| \leq C\, d^s(e + \epsilon X) &= C(1 - |e + \epsilon X|)^s \\ &\leq C(1 - |e + \epsilon X|)^s (1 + |e + \epsilon X|)^s = C(1 - |e + \epsilon X|^2)^s \\ &= C\epsilon^s (-2e \cdot X - \epsilon |X|^2)^s \leq C\epsilon^s, \end{aligned}$$

for some $C > 0$ possibly varying from line to line, and this completes the proof of (4.83).

Now, from (4.83) and the Dominated Convergence Theorem, we obtain

$$\lim_{\epsilon \searrow 0} \int_{\mathbb{R}^n} \epsilon^{-s} \phi_*(e + \epsilon X) \partial^\beta \psi(X)\, dX = \int_{\mathbb{R}^n} \lim_{\epsilon \searrow 0} \epsilon^{-s} \phi_*(e + \epsilon X) \partial^\beta \psi(X)\, dX. \tag{4.84}$$

However, by Corollary 4.14, used here with $\omega := \frac{X}{|X|}$, we know that

$$\begin{aligned} \lim_{\epsilon \searrow 0} \epsilon^{-s} \phi_*(e + \epsilon X) = \lim_{\epsilon \searrow 0} \epsilon^{-s} \phi_*(e + \epsilon |X| \omega) &= |X|^s \lim_{\epsilon \searrow 0} \epsilon^{-s} \phi_*(e + \epsilon \omega) \\ &= k_* |X|^s (-e \cdot \omega)_+^s = k_* (-e \cdot X)_+^s. \end{aligned}$$

Substituting this into (4.84), we obtain

$$\lim_{\epsilon \searrow 0} \int_{\mathbb{R}^n} \epsilon^{-s} \phi_*(e + \epsilon X) \partial^\beta \psi(X)\, dX = k_* \int_{\mathbb{R}^n} (-e \cdot X)_+^s \partial^\beta \psi(X)\, dX.$$

As a consequence, integrating by parts twice,

$$\begin{aligned} &\lim_{\epsilon \searrow 0} \epsilon^{|\beta| - s} \int_{\mathbb{R}^n} \partial^\beta \phi_*(e + \epsilon X) \psi(X)\, dX \\ &\quad = \lim_{\epsilon \searrow 0} \int_{\mathbb{R}^n} \partial^\beta (\epsilon^{-s} \phi_*(e + \epsilon X)) \psi(X)\, dX \\ &\quad = (-1)^{|\beta|} \lim_{\epsilon \searrow 0} \int_{\mathbb{R}^n} \epsilon^{-s} \phi_*(e + \epsilon X) \partial^\beta \psi(X)\, dX \\ &\quad = (-1)^{|\beta|} k_* \int_{\mathbb{R}^n} (-e \cdot X)_+^s \partial^\beta \psi(X)\, dX \\ &\quad = k_* \int_{\mathbb{R}^n} \partial^\beta (-e \cdot X)_+^s \psi(X)\, dX \\ &\quad = (-1)^{|\beta|} k_* s(s-1) \cdots (s - |\beta| + 1) e_1^{\beta_1} \cdots e_n^{\beta_n} \int_{\mathbb{R}^n} (-e \cdot X)_+^{s - |\beta|} \psi(X)\, dX. \end{aligned}$$

Since the test function ψ is arbitrary, the claim in Proposition 4.10 is proved. ☐

4.4 Boundary behavior of s-harmonic functions

In this section we analyze the asymptotic behavior of s-harmonic functions, with a "spherical bump function" as exterior Dirichlet datum.

The result needed for our purpose is the following.

Lemma 4.15. *Let $s > 0$. Let $m \in \mathbb{N}_0$ and $\sigma \in (0,1)$ such that $s = m + \sigma$.*
Then there exists

$$\psi \in H^s(\mathbb{R}^n) \cap C_0^s(\mathbb{R}^n) \quad \text{such that } (-\Delta)^s \psi = 0 \text{ in } B_1, \tag{4.85}$$

and, for every $x \in \partial B_{1-\epsilon}$,

$$\psi(x) = k\epsilon^s + o(\epsilon^s), \tag{4.86}$$

as $\epsilon \searrow 0$, for some $k > 0$.

Proof. Let $\overline{\psi} \in C^\infty(\mathbb{R}, [0,1])$ such that $\overline{\psi} = 0$ in $\mathbb{R} \setminus (2,3)$ and $\overline{\psi} > 0$ in $(2,3)$. Let $\psi_0(x) := (-1)^m \overline{\psi}(|x|)$. We recall the Poisson kernel

$$\Gamma_s(x,y) := (-1)^m \frac{\gamma_{n,\sigma}}{|x-y|^n} \frac{(1-|x|^2)_+^s}{(|y|^2-1)^s},$$

for $x \in \mathbb{R}^n$, $y \in \mathbb{R}^n \setminus \overline{B_1}$, and a suitable normalization constant $\gamma_{n,\sigma} > 0$ (see formulas (1.10) and (1.30) in [3]). We define

$$\psi(x) := \int_{\mathbb{R}^n \setminus B_1} \Gamma_s(x,y)\psi_0(y)\, dy + \psi_0(x).$$

Note that $\psi_0 = 0$ in $B_{3/2}$ and therefore we can exploit Theorem 1.1 in [3] and obtain that (4.85) is satisfied (note also that $\psi = \psi_0$ outside B_1, hence ψ is compactly supported).

Furthermore, to prove (4.86) we borrow some ideas from Lemma 2.2 in [25] and we see that, for any $x \in \partial B_{1-\epsilon}$,

$$\begin{aligned}
\psi(x) &= c(-1)^m \int_{\mathbb{R}^n \setminus B_1} \frac{\psi_0(y)(1-|x|^2)^s}{(|y|^2-1)^s|x-y|^n} dy + \psi_0(x) \\
&= c(-1)^m \int_{\mathbb{R}^n \setminus B_1} \frac{\psi_0(y)(1-|x|^2)^s}{(|y|^2-1)^s|x-y|^n} dy \\
&= c(1-|x|^2)^s \int_2^3 \left[\int_{\mathbb{S}^{n-1}} \frac{\rho^{n-1}\overline{\psi}(\rho)}{(\rho^2-1)^s|x-\rho\omega|^n} d\omega \right] d\rho \\
&= c(2\epsilon-\epsilon^2)^s \int_2^3 \left[\int_{\mathbb{S}^{n-1}} \frac{\rho^{n-1}\overline{\psi}(\rho)}{(\rho^2-1)^s|(1-\epsilon)e_1-\rho\omega|^n} d\omega \right] d\rho
\end{aligned}$$

$$= 2^s c\epsilon^s \int_2^3 \Big[\int_{\mathbb{S}^{n-1}} \frac{\rho^{n-1}\overline{\psi}(\rho)}{(\rho^2-1)^s |e_1 - \rho\omega|^n} d\omega \Big] d\rho + o(\epsilon^s)$$
$$= c\epsilon^s + o(\epsilon^s),$$

where $c > 0$ is a constant possibly varying from line to line, and this establishes (4.86). □

Remark 4.16. As in Proposition 4.10, one can extend (4.86) to higher derivatives (in the distributional sense), obtaining, for any $e \in \partial B_1$ and $\beta \in \mathbb{N}^n$,

$$\lim_{\epsilon \searrow 0} \epsilon^{|\beta|-s} \partial^\beta \psi(e + \epsilon X) = k_\beta e_1^{\beta_1} \cdots e_n^{\beta_n} (-e \cdot X)_+^{s-|\beta|},$$

for some $\kappa_\beta \neq 0$.

Using Lemma 4.15, in the spirit of [25], we can construct a sequence of s-harmonic functions approaching $(x \cdot e)_+^s$ for a fixed unit vector e, by using a blow-up argument. Namely, we prove the following.

Corollary 4.17. *Let $e \in \partial B_1$. There exists a sequence $v_{e,j} \in H^s(\mathbb{R}^n) \cap C^s(\mathbb{R}^n)$ such that $(-\Delta)^s v_{e,j} = 0$ in $B_1(e)$, $v_{e,j} = 0$ in $\mathbb{R}^n \setminus B_{4j}(e)$, and*

$$v_{e,j} \to \kappa (x \cdot e)_+^s \quad \text{in } L^1(B_1(e)),$$

as $j \to +\infty$, for some $\kappa > 0$.

Proof. Let ψ be as in Lemma 4.15 and define

$$v_{e,j}(x) := j^s \psi\Big(\frac{x}{j} - e\Big).$$

The s-harmonicity and the property of being compactly supported follow by the ones of ψ. We now prove the convergence. To this aim, given $x \in B_1(e)$, we write $p_j := \frac{x}{j} - e$ and $\epsilon_j := 1 - |p_j|$. Recall that since $x \in B_1(e)$, we have $|x-e|^2 < 1$, which implies that $|x|^2 < 2x \cdot e$ and $x \cdot e > 0$ for any $x \in B_1(e)$.

As a consequence

$$|p_j|^2 = \Big|\frac{x}{j} - e\Big|^2 = \frac{|x|^2}{j^2} + 1 - 2\frac{x}{j} \cdot e = 1 - \frac{2}{j}(x \cdot e)_+ + o\Big(\frac{1}{j}\Big)(x \cdot e)_+^2,$$

and so

$$\epsilon_j = \frac{(1 + o(1))}{j}(x \cdot e)_+.$$

Therefore, using (4.86),

$$\begin{aligned}v_{e,j}(x) &= j^s\psi(p_j)\\ &= j^s\kappa(\epsilon_j^s + o(\epsilon_j^s))\\ &= j^s\left(\frac{\kappa}{j^s}(x\cdot e)_+^s + o\left(\frac{1}{j^s}\right)\right)\\ &= \kappa(x\cdot e)_+^s + o(1).\end{aligned}$$

Integrating over $B_1(e)$, we obtain the desired L^1-convergence. □

Now, we show that, as in the case $s \in (0,1)$ proved in Theorem 3.1 of [25], we can find an s-harmonic function with an arbitrarily large number of derivatives prescribed at some point.

Proposition 4.18. *For any $\beta \in \mathbb{N}^n$, there exist $p \in \mathbb{R}^n$, $R > r > 0$, and $v \in H^s(\mathbb{R}^n)\cap C^s(\mathbb{R}^n)$ such that*

$$\begin{cases}(-\Delta)^s v = 0 & \text{in } B_r(p),\\ v = 0 & \text{in } \mathbb{R}^n \setminus B_R(p),\end{cases} \tag{4.87}$$

$$\begin{aligned}D^\alpha v(p) = 0 \quad &\text{for any } \alpha \in \mathbb{N}^n \text{ with } |\alpha| \le |\beta| - 1,\\ D^\alpha v(p) = 0 \quad &\text{for any } \alpha \in \mathbb{N}^n \text{ with } |\alpha| = |\beta| \text{ and } \alpha \ne \beta,\end{aligned}$$

and

$$D^\beta v(p) = 1.$$

Proof. Let $\mathcal{Z}$ be the set of all pairs $(v,x) \in (H^s(\mathbb{R}^n) \cap C^s(\mathbb{R}^n)) \times B_r(p)$ that satisfy (4.87) for some $R > r > 0$ and $p \in \mathbb{R}^n$.

To each pair $(v,x) \in \mathcal{Z}$ we associate the vector $(D^\alpha v(x))_{|\alpha|\le|\beta|} \in \mathbb{R}^{K'}$, for some $K' = K'_{|\beta|}$, and consider $\mathcal{V}$ to be the vector space spanned by this construction, namely, we set

$$\mathcal{V} := \{(D^\alpha v(x))_{|\alpha|\le|\beta|}, \text{ with } (v,x) \in \mathcal{Z}\}.$$

We claim that

$$\mathcal{V} = \mathbb{R}^{K'}. \tag{4.88}$$

To check this, we suppose by contradiction that $\mathcal{V}$ lies in a proper subspace of $\mathbb{R}^{K'}$. Then $\mathcal{V}$ must lie in a hyperplane, hence there exists

$$c = (c_\alpha)_{|\alpha|\le|\beta|} \in \mathbb{R}^{K'} \setminus \{0\}, \tag{4.89}$$

which is orthogonal to any vector $(D^\alpha v(x))_{|\alpha|\le|\beta|}$ with $(v,x) \in \mathcal{Z}$, that is,

$$\sum_{|\alpha|\le|\beta|} c_\alpha D^\alpha v(x) = 0. \tag{4.90}$$

We note that the pair $(v_{e,j}, x)$, with v_j as in Corollary 4.17, $e \in \partial B_1$, and $x \in B_1(e)$, belongs to $\mathcal{Z}$. Consequently, for fixed $\xi \in \mathbb{R}^n \setminus B_{1/2}$ and set $e := \frac{\xi}{|\xi|}$, (4.90) holds when $v := v_{e,j}$ and $x \in B_1(e)$, namely,

$$\sum_{|\alpha|\leq|\beta|} c_\alpha D^\alpha v(x) = 0.$$

Let now $\varphi \in C_0^\infty(B_1(e))$. Integrating by parts, by Corollary 4.17 and the Dominated Convergence Theorem, we have

$$\begin{aligned} 0 &= \lim_{j\to+\infty} \int_{\mathbb{R}^n} \sum_{|\alpha|\leq|\beta|} c_\alpha D^\alpha v_{e,j}(x)\varphi(x)\,dx = \lim_{j\to+\infty} \int_{\mathbb{R}^n} \sum_{|\alpha|\leq|\beta|} (-1)^{|\alpha|} c_\alpha v_{e,j}(x) D^\alpha\varphi(x)\,dx \\ &= \kappa \int_{\mathbb{R}^n} \sum_{|\alpha|\leq|\beta|} (-1)^{|\alpha|} c_\alpha (x\cdot e)_+^s D^\alpha \varphi(x)\,dx = \kappa \int_{\mathbb{R}^n} \sum_{|\alpha|\leq|\beta|} c_\alpha D^\alpha (x\cdot e)_+^s \varphi(x)\,dx. \end{aligned}$$

This gives, for every $x \in B_1(e)$,

$$\sum_{|\alpha|\leq|\beta|} c_\alpha D^\alpha (x\cdot e)_+^s = 0.$$

Moreover, for every $x \in B_1(e)$,

$$D^\alpha (x\cdot e)_+^s = s(s-1)\cdots(s-|\alpha|+1)(x\cdot e)_+^{s-|\alpha|} e_1^{\alpha_1}\cdots e_n^{\alpha_n}.$$

In particular, for $x = \frac{e}{|\xi|} \in B_1(e)$,

$$D^\alpha (x\cdot e)_+^s \big|_{x=e/|\xi|} = s(s-1)\cdots(s-|\alpha|+1)|\xi|^{-s}\xi_1^{\alpha_1}\cdots\xi_n^{\alpha_n}.$$

And, using the usual multi-index notation, we write

$$\sum_{|\alpha|\leq|\beta|} c_\alpha s(s-1)\cdots(s-|\alpha|+1)\xi^\alpha = 0, \tag{4.91}$$

for any $\xi \in \mathbb{R}^n \setminus B_{1/2}$. The identity (4.91) describes a polynomial in ξ which vanishes for any ξ in an open subset of $\mathbb{R}^n$. As a result, the Identity Principle for polynomials leads to

$$c_\alpha s(s-1)\cdots(s-|\alpha|+1) = 0,$$

for all $|\alpha| \leq |\beta|$.

Consequently, since $s \in \mathbb{R}\setminus\mathbb{N}$, the product $s(s-1)\cdots(s-|\alpha|+1)$ never vanishes, and so the coefficients c_α are forced to be null for any $|\alpha| \leq |\beta|$. This is in contradiction with (4.89), and therefore the proof of (4.88) is complete.

From this, the desired claim in Proposition 4.18 plainly follows. □

5 Proof of the main result

This chapter is devoted to the proof of the main result in Theorem 2.1. This will be accomplished by an auxiliary result of purely nonlocal type which will allow us to prescribe an arbitrarily large number of derivatives at a point for the solution of a fractional equation.

5.1 A result which implies Theorem 2.1

We will use the notation

$$\Lambda_{-\infty} := \Lambda_{(-\infty,\dots,-\infty)}, \tag{5.1}$$

that is, we exploit (2.8) with $a_1 := \cdots := a_l := -\infty$. In this section we present the following statement.

Theorem 5.1. *Suppose that*

$$\begin{aligned}&\textit{either there exists } i \in \{1,\dots,M\} \textit{ such that } \mathfrak{b}_i \neq 0 \textit{ and } s_i \notin \mathbb{N},\\ &\textit{or there exists } i \in \{1,\dots,l\} \textit{ such that } \mathfrak{c}_i \neq 0 \textit{ and } \alpha_i \notin \mathbb{N}.\end{aligned}$$

Let $\ell \in \mathbb{N}$, $f : \mathbb{R}^N \to \mathbb{R}$, with $f \in C^\ell(\overline{B_1^N})$. For fixed $\epsilon > 0$, there exist

$$\begin{aligned} &u = u_\epsilon \in C^\infty(B_1^N) \cap C(\mathbb{R}^N),\\ &a = (a_1,\dots,a_l) = (a_{1,\epsilon},\dots,a_{l,\epsilon}) \in (-\infty,0)^l,\\ \textit{and}\quad &R = R_\epsilon > 1\end{aligned}$$

such that:

- *for every $h \in \{1,\dots,l\}$ and $(x,y,t_1,\dots,t_{h-1},t_{h+1},\dots,t_l)$*

$$\textit{the map } \mathbb{R} \ni t_h \mapsto u(x,y,t) \textit{ belongs to } C_{-\infty}^{k_h,\alpha_h}, \tag{5.2}$$

  in the notation of formula (1.4) of [20]*;*
- *we have*

$$\begin{cases} \Lambda_{-\infty}u = 0 & \textit{in } B_1^{N-l} \times (-1,+\infty)^l,\\ u(x,y,t) = 0 & \textit{if } |(x,y)| \geq R,\end{cases} \tag{5.3}$$

$$\partial_{t_h}^{k_h} u(x,y,t) = 0 \quad \textit{if } t_h \in (-\infty,a_h), \textit{ for all } h \in \{1,\dots,l\}, \tag{5.4}$$

  and

$$\|u - f\|_{C^\ell(B_1^N)} < \epsilon. \tag{5.5}$$

https://doi.org/10.1515/9783110664355-005

The proof of Theorem 5.1 will basically occupy the rest of this work, and this will lead us to the completion of the proof of Theorem 2.1. Indeed, we have the following.

Lemma 5.2. *If the statement of Theorem 5.1 holds, then the statement of Theorem 2.1 holds.*

Proof. Assume that the claims in Theorem 5.1 are satisfied. Then, by (5.2) and (5.4), we are in the position of exploiting Lemma A.1 in [20] and conclude that, in $B_1^{N-l} \times (-1,+\infty)^l$,

$$D^{\alpha_h}_{t_h,a_h} u = D^{\alpha_h}_{t_h,-\infty} u,$$

for every $h \in \{1,\dots,l\}$. This and (5.3) give

$$\Lambda_a u = \Lambda_{-\infty} u = 0 \quad \text{in } B_1^{N-l} \times (-1,+\infty)^l. \tag{5.6}$$

We also define

$$\underline{a} := \min_{h\in\{1,\dots,l\}} a_h$$

and take $\tau \in C_0^\infty([-\underline{a}-2,3])$ with $\tau = 1$ in $[-\underline{a}-1,1]$. Let

$$U(x,y,t) := u(x,y,t)\tau(t_1)\cdots\tau(t_l). \tag{5.7}$$

Our goal is to prove that U satisfies the theses of Theorem 2.1. To this end, we observe that $u = U$ in B_1^N, and therefore (2.12) for U plainly follows from (5.5).

In addition, from (2.6), we see that $D^{\alpha_h}_{t_h,a_h}$ at a point $t_h \in (-1,1)$ only depends on the values of the function between a_h and 1. Since the cutoffs in (5.7) do not alter these values, we see that $D^{\alpha_h}_{t_h,a_h} U = D^{\alpha_h}_{t_h,a_h} u$ in B_1^N, and accordingly $\Lambda_a U = \Lambda_a u$ in B_1^N. Form this and (5.6) it follows that

$$\Lambda_a U = 0 \quad \text{in } B_1^N. \tag{5.8}$$

Also, since u in Theorem 5.1 is compactly supported in the variable (x,y), we see from (5.7) that U is compactly supported in the variables (x,y,t). From this and (5.8) it follows that (2.11) is satisfied by U (up to renaming R). □

5.2 A pivotal span result towards the proof of Theorem 5.1

In what follows, we let $\Lambda_{-\infty}$ be as in (5.1), we recall the setting in (2.1), and we use the following multi-index notations:

$$\begin{aligned} \iota &= (i, I, \mathfrak{I}) = (i_1,\dots,i_n,I_1,\dots,I_M,\mathfrak{I}_1,\dots,\mathfrak{I}_l) \in \mathbb{N}^N \\ \text{and} \quad \partial^\iota w &= \partial^{i_1}_{x_1}\cdots\partial^{i_n}_{x_n}\partial^{I_1}_{y_1}\cdots\partial^{I_M}_{y_M}\partial^{\mathfrak{I}_1}_{t_1}\cdots\partial^{\mathfrak{I}_l}_{t_l} w. \end{aligned} \tag{5.9}$$

Inspired by Lemma 5 of [26], we consider the span of the derivatives of functions in $\ker \Lambda_{-\infty}$, with derivatives up to a fixed order $K \in \mathbb{N}$. We want to prove that the derivatives of such functions span a maximal vectorial space.

For this, we denote by $\partial^K w(0)$ the vector with entries given, in some prescribed order, by $\partial^\iota w(0)$ with $|\iota| \le K$.

We note that

$$\partial^K w(0) \in \mathbb{R}^{K'} \quad \text{for some } K' \in \mathbb{N}, \tag{5.10}$$

with K' depending on K.

Now, we adopt the notation in formula (1.4) of [20], and we denote by $\mathcal{A}$ the set of all functions $w = w(x, y, t)$ such that for all $h \in \{1, \dots, l\}$ and all $(x, y, t_1, \dots, t_{h-1}, t_{h+1}, \dots, t_l) \in \mathbb{R}^{N-1}$, the map $\mathbb{R} \ni t_h \mapsto w(x, y, t)$ belongs to $C^\infty((a_h, +\infty)) \cap C^{k_h,\alpha_h}_{-\infty}$, and (5.4) holds for some $a_h \in (-2, 0)$.

We also set

$$\mathcal{H} := \{w \in C(\mathbb{R}^N) \cap C_0(\mathbb{R}^{N-l}) \cap C^\infty(\mathcal{N}) \cap \mathcal{A}, \text{ for some neighborhood } \mathcal{N} \text{ of the origin,}$$
$$\text{such that } \Lambda_{-\infty} w = 0 \text{ in } \mathcal{N}\}$$

and, for any $w \in \mathcal{H}$, we let $\mathcal{V}_K$ be the vector space spanned by the vector $\partial^K w(0)$.

By (5.10), we know that $\mathcal{V}_K \subseteq \mathbb{R}^{K'}$. In fact, we show that equality holds in this inclusion, as stated in the following[1] result.

Lemma 5.3. *we have* $\mathcal{V}_K = \mathbb{R}^{K'}$.

The proof of Lemma 5.3 is by contradiction. Namely, if $\mathcal{V}_K$ does not exhaust the whole of $\mathbb{R}^{K'}$ there exists

$$\theta \in \partial B_1^{K'} \tag{5.11}$$

such that

$$\mathcal{V}_K \subseteq \{\zeta \in \mathbb{R}^{K'} \text{ s.t. } \theta \cdot \zeta = 0\}. \tag{5.12}$$

In coordinates, recalling (5.9), we write θ as $\theta_\iota = \theta_{i,I,\mathfrak{J}}$, with $i \in \mathbb{N}^{p_1+\dots+p_n}$, $I \in \mathbb{N}^{m_1+\dots+m_M}$, and $\mathfrak{J} \in \mathbb{N}^l$. We consider

$$\begin{aligned}&\text{a multi-index } \bar{I} \in \mathbb{N}^{m_1+\dots+m_M} \text{ such that it maximizes } |I|\\ &\text{among all the multi-indices } (i, I, \mathfrak{J}) \text{ for which } |i| + |I| + |\mathfrak{J}| \le K\\ &\text{and } \theta_{i,I,\mathfrak{J}} \neq 0 \text{ for some } (i, \mathfrak{J}).\end{aligned} \tag{5.13}$$

1 Note that results analogous to Lemma 5.3 cannot hold for solutions of local operators. For instance, pure second derivatives of harmonic functions have to satisfy a linear equation, so they are forced to lie in a proper subspace. In this sense, results such as Lemma 5.3 here reveal a truly nonlocal phenomenon.

Some comments on the setting in (5.13) follow here. We stress that, by (5.11), the set $\mathcal{S}$ of indices I for which there exist indices $(i, \mathfrak{I})$ such that $|i| + |I| + |\mathfrak{I}| \le K$ and $\theta_{i,I,\mathfrak{I}} \neq 0$ is not empty. Therefore, since $\mathcal{S}$ is a finite set, we can take

$$S := \sup_{I \in \mathcal{S}} |I| = \max_{I \in \mathcal{S}} |I| \in \mathbb{N} \cap [0, K].$$

Hence, we consider a multi-index $\bar{I}$ for which $|\bar{I}| = S$ to obtain the setting in (5.13). By construction, we have

- $|i| + |\bar{I}| + |\mathfrak{I}| \le K$;
- if $|I| > |\bar{I}|$, then $\theta_{i,I,\mathfrak{I}} = 0$;
- and there exist multi-indices i and $\mathfrak{I}$ such that $\theta_{i,\bar{I},\mathfrak{I}} \neq 0$.

As a variation of the setting in (5.13), we can also consider

$$\begin{aligned}&\text{a multi-index } \bar{\mathfrak{I}} \in \mathbb{N}^l \text{ such that it maximizes } |\mathfrak{I}| \\ &\text{among all multi-indices } (i, I, \mathfrak{I}) \text{ for which } |i| + |I| + |\mathfrak{I}| \le K \\ &\text{and } \theta_{i,I,\mathfrak{I}} \neq 0 \text{ for some } (i, I).\end{aligned} \tag{5.14}$$

In the setting of (5.13) and (5.14), we claim that there exists an open set of $\mathbb{R}^{p_1+\cdots+p_n} \times \mathbb{R}^{m_1+\cdots+m_M} \times \mathbb{R}^l$ such that for every $(\underline{x}, \underline{y}, \underline{t})$ in such open set we have

$$\begin{aligned}\text{either} \quad 0 &= \sum_{\substack{|i|+|I|+|\mathfrak{I}|\le K \\ |I|=|\bar{I}|}} C_{i,I,\mathfrak{I}} \theta_{i,I,\mathfrak{I}} \underline{x}^i \underline{y}^I \underline{t}^{\mathfrak{I}}, \quad \text{with } C_{i,I,\mathfrak{I}} \neq 0, \\ \text{or} \quad 0 &= \sum_{\substack{|i|+|I|+|\mathfrak{I}|\le K \\ |\mathfrak{I}|=|\bar{\mathfrak{I}}|}} C_{i,I,\mathfrak{I}} \theta_{i,I,\mathfrak{I}} \underline{x}^i \underline{y}^I \underline{t}^{\mathfrak{I}}, \quad \text{with } C_{i,I,\mathfrak{I}} \neq 0.\end{aligned} \tag{5.15}$$

In our framework, the claim in (5.15) will be pivotal towards the completion of the proof of Lemma 5.3. Indeed, let us suppose for the moment that (5.15) is established and let us complete the proof of Lemma 5.3 by arguing as follows.

Formula (5.15) says that $\theta \cdot \partial^K w(0)$ is a polynomial which vanishes for any triple $(\underline{x}, \underline{y}, \underline{t})$ in an open subset of $\mathbb{R}^{p_1+\cdots+p_n} \times \mathbb{R}^{m_1+\cdots+m_M} \times \mathbb{R}^l$. Hence, using the Identity Principle for polynomials, each $C_{i,I,\mathfrak{I}}\theta_{i,I,\mathfrak{I}}$ is equal to zero whenever $|i| + |I| + |\mathfrak{I}| \le K$ and either $|I| = |\bar{I}|$ (if the first identity in (5.15) holds) or $|\mathfrak{I}| = |\bar{\mathfrak{I}}|$ (if the second identity in (5.15) holds). Then, since $C_{i,I,\mathfrak{I}} \neq 0$, we conclude that each $\theta_{i,I,\mathfrak{I}}$ is zero as long as either $|I| = |\bar{I}|$ (in the first case) or $|\mathfrak{I}| = |\bar{\mathfrak{I}}|$ (in the second case), but this contradicts either the definition of $\bar{I}$ in (5.13) (in the first case) or the definition of $\bar{\mathfrak{I}}$ in (5.14) (in the second case). This would therefore complete the proof of Lemma 5.3.

In view of the discussion above, it remains to prove (5.15). To this end, we distinguish the following four cases:

1. there exist $i \in \{1, \dots, n\}$ and $j \in \{1, \dots, M\}$ such that $\mathfrak{a}_i \neq 0$ and $\mathfrak{b}_j \neq 0$;
2. there exist $i \in \{1, \dots, n\}$ and $h \in \{1, \dots, l\}$ such that $\mathfrak{a}_i \neq 0$ and $\mathfrak{c}_h \neq 0$;

3. we have $\mathfrak{a}_1 = \cdots = \mathfrak{a}_n = 0$, and there exists $j \in \{1, \ldots, M\}$ such that $\mathfrak{b}_j \neq 0$;
4. we have $\mathfrak{a}_1 = \cdots = \mathfrak{a}_n = 0$, and there exists $h \in \{1, \ldots, l\}$ such that $\mathfrak{c}_h \neq 0$.

Note that cases 1 and 3 deal with the case in which space-fractional diffusion is present (and in case 1 one also has classical derivatives, while in case 3 the classical derivatives are absent).

Similarly, cases 2 and 4 deal with the case in which time-fractional diffusion is present (and in case 2 one also has classical derivatives, while in case 4 the classical derivatives are absent).

Of course, the case in which both space- and time-fractional diffusion occurs is already comprised by the previous cases (namely, it is comprised in both cases 1 and 2 if classical derivatives are also present, and in both cases 3 and 4 if classical derivatives are absent).

Proof of (5.15), *case 1.* For any $j \in \{1, \ldots, M\}$ we denote by $\tilde{\phi}_{\star,j}$ the first eigenfunction for $(-\Delta)^{s_j}_{y_j}$ vanishing outside $B_1^{m_j}$ given in Corollary 4.9. We normalize it such that $\|\tilde{\phi}_{\star,j}\|_{L^2(\mathbb{R}^{m_j})} = 1$, and we write $\lambda_{\star,j} \in (0, +\infty)$ to indicate the corresponding first eigenvalue (which now depends on s_j), namely, we write

$$\begin{cases} (-\Delta)^{s_j}_{y_j} \tilde{\phi}_{\star,j} = \lambda_{\star,j} \tilde{\phi}_{\star,j} & \text{in } B_1^{m_j}, \\ \tilde{\phi}_{\star,j} = 0 & \text{in } \mathbb{R}^{m_j} \setminus \overline{B_1^{m_j}}. \end{cases} \tag{5.16}$$

Up to reordering the variables and/or taking the operators to the other side of the equation, given the assumptions of case 1, we suppose that

$$\mathfrak{a}_1 \neq 0 \tag{5.17}$$

and

$$\mathfrak{b}_M > 0. \tag{5.18}$$

In view of (5.17), we define

$$R := \left(\frac{1}{|\mathfrak{a}_1|} \left(\sum_{j=1}^{M-1} |\mathfrak{b}_j| \lambda_{\star,j} + \sum_{h=1}^{l} |\mathfrak{c}_h| \right) \right)^{1/|r_1|}. \tag{5.19}$$

Now, we fix two sets of free parameters

$$\underline{x}_1 \in (R+1, R+2)^{p_1}, \ldots, \underline{x}_n \in (R+1, R+2)^{p_n}, \tag{5.20}$$

and

$$\underline{t}_{\star,1} \in \left(\frac{1}{2}, 1\right), \ldots, \underline{t}_{\star,l} \in \left(\frac{1}{2}, 1\right). \tag{5.21}$$

We also set

$$\lambda_j := \lambda_{\star,j} \quad \text{for } j \in \{1,\dots,M-1\}, \tag{5.22}$$

where $\lambda_{\star,j}$ is defined as in (5.16), and

$$\lambda_M := \frac{1}{\mathfrak{b}_M}\left(\sum_{j=1}^{n} |\mathfrak{a}_j| \underline{\mathfrak{x}}_j^{r_j} - \sum_{j=1}^{M-1} \mathfrak{b}_j \lambda_j - \sum_{h=1}^{l} \mathfrak{c}_h \underline{\mathfrak{t}}_{\star,h}\right). \tag{5.23}$$

Note that this definition is well-posed, thanks to (5.18). In addition, by (5.20), we can write $\underline{\mathfrak{x}}_j = (\underline{\mathfrak{x}}_{j1},\dots,\underline{\mathfrak{x}}_{jp_j})$, and we know that $\underline{\mathfrak{x}}_{j\ell} > R+1$ for any $j \in \{1,\dots,n\}$ and any $\ell \in \{1,\dots,p_j\}$. Therefore,

$$\underline{\mathfrak{x}}_j^{r_j} = \underline{\mathfrak{x}}_{j1}^{r_{j1}} \cdots \underline{\mathfrak{x}}_{jp_j}^{r_{jp_j}} \geq 0. \tag{5.24}$$

From this, (5.19), and (5.21), we deduce that

$$\begin{aligned}
\sum_{j=1}^{n} |\mathfrak{a}_j| \underline{\mathfrak{x}}_j^{r_j} &\geq |\mathfrak{a}_1| \underline{\mathfrak{x}}_1^{r_1} \geq |\mathfrak{a}_1| (R+1)^{|r_1|} > |\mathfrak{a}_1| R^{|r_1|} \\
&= \sum_{j=1}^{M-1} |\mathfrak{b}_j| \lambda_j + \sum_{h=1}^{l} |\mathfrak{c}_h| \geq \sum_{j=1}^{M-1} \mathfrak{b}_j \lambda_j + \sum_{h=1}^{l} \mathfrak{c}_h \underline{\mathfrak{t}}_{\star,h},
\end{aligned}$$

and consequently, by (5.23),

$$\lambda_M > 0. \tag{5.25}$$

We also set

$$\omega_j := \begin{cases} 1 & \text{if } j = 1,\dots,M-1, \\ \dfrac{\lambda_{\star,M}^{1/2s_M}}{\lambda_M^{1/2s_M}} & \text{if } j = M. \end{cases} \tag{5.26}$$

Note that this definition is well-posed, thanks to (5.25). In addition, by (5.16), for any $j \in \{1,\dots,M\}$, the functions

$$\phi_j(y_j) := \tilde{\phi}_{\star,j}\left(\frac{y_j}{\omega_j}\right) \tag{5.27}$$

are eigenfunctions of $(-\Delta)_{y_j}^{s_j}$ in $B_{\omega_j}^{m_j}$ with external homogenous Dirichlet boundary condition and eigenvalues λ_j; namely, we can rewrite (5.16) as

$$\begin{cases} (-\Delta)_{y_j}^{s_j} \phi_j = \lambda_j \phi_j & \text{in } B_{\omega_j}^{m_j}, \\ \phi_j = 0 & \text{in } \mathbb{R}^{m_j} \setminus \overline{B_{\omega_j}^{m_j}}. \end{cases} \tag{5.28}$$

Now, we define

$$\psi_{\star,h}(t_h) := E_{\alpha_h,1}(t_h^{\alpha_h}), \tag{5.29}$$

where $E_{\alpha_h,1}$ denotes the Mittag-Leffler function with parameters $\alpha := \alpha_h$ and $\beta := 1$ as defined in (3.1).

Moreover, we consider $a_h \in (-2, 0)$, for every $h = 1, \dots, l$, to be chosen appropriately in what follows (the precise choice will be performed in (5.48)), and, recalling (5.21), we let

$$\underline{t}_h := \underline{t}_{\star,h}^{1/\alpha_h}, \tag{5.30}$$

and we define

$$\psi_h(t_h) := \psi_{\star,h}(\underline{t}_h(t_h - a_h)) = E_{\alpha_h,1}(\underline{t}_{\star,h}(t_h - a_h)^{\alpha_h}). \tag{5.31}$$

We point out that, thanks to Lemma 3.2, the function in (5.31) solves

$$\begin{cases} D^{\alpha_h}_{t_h,a_h}\psi_h(t_h) = \underline{t}_{\star,h}\psi_h(t_h) & \text{in } (a_h, +\infty), \\ \psi_h(a_h) = 1, \\ \partial^m_{t_h}\psi_h(a_h) = 0 & \text{for every } m \in \{1, \dots, [\alpha_h]\}. \end{cases} \tag{5.32}$$

Moreover, for any $h \in \{1, \dots, l\}$, we define

$$\psi_h^\star(t_h) := \begin{cases} \psi_h(t_h) & \text{if } t_h \in [a_h, +\infty), \\ 1 & \text{if } t_h \in (-\infty, a_h). \end{cases} \tag{5.33}$$

Thanks to (5.32) and Lemma A.3 in [20] applied here with $b := a_h$, $a := -\infty$, $u := \psi_h$, $u_\star := \psi_h^\star$, we have $\psi_h^\star \in C^{k_h,\alpha_h}_{-\infty}$, and

$$\begin{aligned} &D^{\alpha_h}_{t_h,-\infty}\psi_h^\star(t_h) = D^{\alpha_h}_{t_h,a_h}\psi_h(t_h) = \underline{t}_{\star,h}\psi_h(t_h) = \underline{t}_{\star,h}\psi_h^\star(t_h) \\ &\quad \text{in every interval } I \Subset (a_h, +\infty). \end{aligned} \tag{5.34}$$

We observe that the setting in (5.33) is compatible with the ones in (5.2) and (5.4).

From (3.1) and (5.31), we see that

$$\psi_h(t_h) = \sum_{j=0}^{+\infty} \frac{\underline{t}^j_{\star,h}(t_h - a_h)^{\alpha_h j}}{\Gamma(\alpha_h j + 1)}.$$

Consequently, for every $\mathfrak{I}_h \in \mathbb{N}$, we have

$$\partial^{\mathfrak{I}_h}_{t_h}\psi_h(t_h) = \sum_{j=0}^{+\infty} \frac{\underline{t}^j_{\star,h}\alpha_h j(\alpha_h j - 1)\dots(\alpha_h j - \mathfrak{I}_h + 1)(t_h - a_h)^{\alpha_h j - \mathfrak{I}_h}}{\Gamma(\alpha_h j + 1)}. \tag{5.35}$$

Now, we define, for any $i \in \{1, \dots, n\}$,

$$\overline{a}_i := \begin{cases} \frac{a_i}{|a_i|} & \text{if } a_i \neq 0, \\ 1 & \text{if } a_i = 0. \end{cases}$$

We note that

$$\overline{a}_i \neq 0 \quad \text{for all } i \in \{1, \dots, n\}, \tag{5.36}$$

and

$$a_i \overline{a}_i = |a_i|. \tag{5.37}$$

Now, for each $i \in \{1, \dots, n\}$, we consider the multi-index $r_i = (r_{i1}, \dots, r_{ip_i}) \in \mathbb{N}^{p_i}$. This multi-index acts on $\mathbb{R}^{p_i}$, whose variables are denoted by $x_i = (x_{i1}, \dots, x_{ip_i}) \in \mathbb{R}^{p_i}$. We let $\overline{v}_{i1}$ be the solution of the Cauchy problem

$$\begin{cases} \partial_{x_{i1}}^{r_{i1}} \overline{v}_{i1} = -\overline{a}_i \overline{v}_{i1}, \\ \partial_{x_{i1}}^{\beta_1} \overline{v}_{i1}(0) = 1 \quad \text{for every } \beta_1 \leq r_{i1} - 1. \end{cases} \tag{5.38}$$

We note that the solution of the Cauchy problem in (5.38) exists at least in a neighborhood of the origin of the form $[-\rho_{i1}, \rho_{i1}]$ for a suitable $\rho_{i1} > 0$.

Moreover, if $p_i \geq 2$, for any $\ell \in \{2, \dots, p_i\}$, we consider the solution of the following Cauchy problem:

$$\begin{cases} \partial_{x_{i\ell}}^{r_{i\ell}} \overline{v}_{i\ell} = \overline{v}_{i\ell}, \\ \partial_{x_{i\ell}}^{\beta_\ell} \overline{v}_{i\ell}(0) = 1 \quad \text{for every } \beta_\ell \leq r_{i\ell} - 1. \end{cases} \tag{5.39}$$

As above, these solutions are well-defined at least in a neighborhood of the origin of the form $[-\rho_{i\ell}, \rho_{i\ell}]$, for a suitable $\rho_{i\ell} > 0$.

Then, we define

$$\overline{\rho}_i := \min\{\rho_{i1}, \dots, \rho_{ip_i}\} = \min_{\ell \in \{1, \dots, p_i\}} \rho_{i\ell}.$$

In this way, for every $x_i = (x_{i1}, \dots, x_{ip_i}) \in B_{\overline{\rho}_i}^{p_i}$, we set

$$\overline{v}_i(x_i) := \overline{v}_{i1}(x_{i1}) \cdots \overline{v}_{ip_i}(x_{ip_i}). \tag{5.40}$$

By (5.38) and (5.39), we have

$$\begin{cases} \partial_{x_i}^{r_i} \overline{v}_i = -\overline{a}_i \overline{v}_i, \\ \partial_{x_i}^{\beta} \overline{v}_i(0) = 1 & \begin{array}{l} \text{for every } \beta = (\beta_1, \dots \beta_{p_i}) \in \mathbb{N}^{p_i} \\ \text{such that } \beta_\ell \leq r_{i\ell} - 1 \text{ for each } \ell \in \{1, \dots, p_i\}. \end{array} \end{cases} \tag{5.41}$$

Now, we define

$$\rho := \min\{\overline{\rho}_1, \dots \overline{\rho}_n\} = \min_{i\in\{1,\dots,n\}} \overline{\rho}_i.$$

We take

$$\overline{\tau} \in C_0^\infty(B^{p_1+\dots+p_n}_{\rho/(R+2)}),$$

with $\overline{\tau} = 1$ in $B^{p_1+\dots+p_n}_{\rho/(2(R+2))}$, and, for every $x = (x_1, \dots, x_n) \in \mathbb{R}^{p_1} \times \dots \times \mathbb{R}^{p_n}$, we set

$$\tau_1(x_1, \dots, x_n) := \overline{\tau}(\underline{\mathcal{X}}_1 \otimes x_1, \dots, \underline{\mathcal{X}}_n \otimes x_n). \tag{5.42}$$

We recall that the free parameters $\underline{\mathcal{X}}_1, \dots, \underline{\mathcal{X}}_n$ have been introduced in (5.20), and we have used here the notation

$$\underline{\mathcal{X}}_i \otimes x_i = (\underline{\mathcal{X}}_{i1}, \dots, \underline{\mathcal{X}}_{ip_i}) \otimes (x_{i1}, \dots, x_{ip_i}) := (\underline{\mathcal{X}}_{i1}x_{i1}, \dots, \underline{\mathcal{X}}_{ip_i}x_{ip_i}) \in \mathbb{R}^{p_i},$$

for every $i \in \{1, \dots, n\}$.

We also set, for any $i \in \{1, \dots, n\}$,

$$v_i(x_i) := \overline{v}_i(\underline{\mathcal{X}}_i \otimes x_i). \tag{5.43}$$

We point out that if $x_i \in B^{p_i}_{\overline{\rho}_i/(R+2)}$ we have

$$|\underline{\mathcal{X}}_i \otimes x_i|^2 = \sum_{\ell=1}^{p_i} (\underline{\mathcal{X}}_{i\ell} x_{i\ell})^2 \le (R+2)^2 \sum_{\ell=1}^{p_i} x_{i\ell}^2 < \overline{\rho}_i^2,$$

thanks to (5.20), and therefore the setting in (5.43) is well-defined for every $x_i \in B^{p_i}_{\overline{\rho}_i/(R+2)}$.

Recalling (5.41) and (5.43), we see that, for any $i \in \{1, \dots, n\}$,

$$\partial^{r_i}_{x_i} v_i(x_i) = \underline{\mathcal{X}}_i^{r_i} \partial^{r_i}_{x_i} \overline{v}_i(\underline{\mathcal{X}}_i \otimes x_i) = -\overline{a}_i \underline{\mathcal{X}}_i^{r_i} \overline{v}_i(\underline{\mathcal{X}}_i \otimes x_i) = -\overline{a}_i \underline{\mathcal{X}}_i^{r_i} v_i(x_i). \tag{5.44}$$

We take $e_1, \dots, e_M$, with

$$e_j \in \partial B^{m_j}_{\omega_j}, \tag{5.45}$$

and we introduce an additional set of free parameters $Y_1, \dots, Y_M$ with

$$Y_j \in \mathbb{R}^{m_j} \quad \text{and} \quad e_j \cdot Y_j < 0. \tag{5.46}$$

We let $\epsilon > 0$, to be taken small and possibly depending on the free parameters e_j, Y_j, and $\underline{t}_h$, and we define

$$\begin{aligned} w(x, y, t) := {} & \tau_1(x) v_1(x_1) \cdots v_n(x_n) \phi_1(y_1 + e_1 + \epsilon Y_1) \cdots \phi_M(y_M + e_M + \epsilon Y_M) \\ & \times \psi_1^\star(t_1) \cdots \psi_l^\star(t_l), \end{aligned} \tag{5.47}$$

where the setting in (5.27), (5.33), (5.42), and (5.43) has been exploited.

We also note that $w \in C(\mathbb{R}^N) \cap C_0(\mathbb{R}^{N-l}) \cap \mathcal{A}$. Moreover, if

$$a = (a_1, \dots, a_l) := \left(-\frac{\epsilon}{\underline{\mathfrak{t}}_1}, \dots, -\frac{\epsilon}{\underline{\mathfrak{t}}_l} \right) \in (-\infty, 0)^l, \tag{5.48}$$

(x, y) is sufficiently close to the origin, and $t \in (a_1, +\infty) \times \cdots \times (a_l, +\infty)$, we have

$$\begin{aligned}
&\Lambda_{-\infty} w(x,y,t)\\
&= \left(\sum_{i=1}^{n} \mathfrak{a}_i \partial_{x_i}^{r_i} + \sum_{j=1}^{M} \mathfrak{b}_j (-\Delta)_{y_j}^{s_j} + \sum_{h=1}^{l} \mathfrak{c}_h D_{t_h,-\infty}^{\alpha_h} \right) w(x,y,t)\\
&= \sum_{i=1}^{n} \mathfrak{a}_i v_1(x_1) \cdots v_{i-1}(x_{i-1}) \partial_{x_i}^{r_i} v_i(x_i) v_{i+1}(x_{i+1}) \cdots v_n(x_n)\\
&\quad \times \phi_1(y_1 + e_1 + \epsilon Y_1) \cdots \phi_M(y_M + e_M + \epsilon Y_M) \psi_1^\star(t_1) \cdots \psi_l^\star(t_l)\\
&\quad + \sum_{j=1}^{M} \mathfrak{b}_j v_1(x_1) \cdots v_n(x_n) \phi_1(y_1 + e_1 + \epsilon Y_1) \cdots \phi_{j-1}(y_{j-1} + e_{j-1} + \epsilon Y_{j-1})\\
&\quad \times (-\Delta)_{y_j}^{s_j} \phi_j(y_j + e_j + \epsilon Y_j) \phi_{j+1}(y_{j+1} + e_{j+1} + \epsilon Y_{j+1}) \cdots \phi_M(y_M + e_M + \epsilon Y_M)\\
&\quad \times \psi_1^\star(t_1) \cdots \psi_l^\star(t_l)\\
&\quad + \sum_{h=1}^{l} \mathfrak{c}_h v_1(x_1) \cdots v_n(x_n) \phi_1(y_1 + e_1 + \epsilon Y_1) \cdots \phi_M(y_M + e_M + \epsilon Y_M) \psi_1^\star(t_1) \cdots \psi_{h-1}^\star(t_{h-1})\\
&\quad \times D_{t_h,-\infty}^{\alpha_h} \psi_h^\star(t_h) \psi_{h+1}^\star(t_{h+1}) \cdots \psi_l^\star(t_l)\\
&= -\sum_{i=1}^{n} \mathfrak{a}_i \bar{\mathfrak{a}}_i \underline{\mathfrak{x}}_i^{r_i} v_1(x_1) \cdots v_n(x_n) \phi_1(y_1 + e_1 + \epsilon Y_1) \cdots \phi_M(y_M + e_M + \epsilon Y_M) \psi_1^\star(t_1) \cdots \psi_l^\star(t_l)\\
&\quad + \sum_{j=1}^{M} \mathfrak{b}_j \lambda_j v_1(x_1) \cdots v_n(x_n) \phi_1(y_1 + e_1 + \epsilon Y_1) \cdots \phi_M(y_M + e_M + \epsilon Y_M) \psi_1^\star(t_1) \cdots \psi_l^\star(t_l)\\
&\quad + \sum_{h=1}^{l} \mathfrak{c}_h \underline{\mathfrak{t}}_{\star,h} v_1(x_1) \cdots v_n(x_n) \phi_1(y_1 + e_1 + \epsilon Y_1) \cdots \phi_M(y_M + e_M + \epsilon Y_M) \psi_1^\star(t_1) \cdots \psi_l^\star(t_l)\\
&= \left(-\sum_{i=1}^{n} \mathfrak{a}_i \bar{\mathfrak{a}}_i \underline{\mathfrak{x}}_i^{r_i} + \sum_{j=1}^{M} \mathfrak{b}_j \lambda_j + \sum_{h=1}^{l} \mathfrak{c}_h \underline{\mathfrak{t}}_{\star,h} \right) w(x,y,t),
\end{aligned}$$

thanks to (5.28), (5.34), and (5.44).

Consequently, making use of (5.22), (5.23), and (5.37), if (x, y) lies near the origin and $t \in (a_1, +\infty) \times \cdots \times (a_l, +\infty)$, we have

$$\begin{aligned}
\Lambda_{-\infty} w(x,y,t) &= \left(-\sum_{i=1}^{n} |\mathfrak{a}_i| \underline{\mathfrak{x}}_i^{r_i} + \sum_{j=1}^{M-1} \mathfrak{b}_j \lambda_j + \mathfrak{b}_M \lambda_M + \sum_{h=1}^{l} \mathfrak{c}_h \underline{\mathfrak{t}}_{\star,h} \right) w(x,y,t)\\
&= \left(-\sum_{i=1}^{n} |\mathfrak{a}_i| \underline{\mathfrak{x}}_i^{r_i} + \sum_{j=1}^{M-1} \mathfrak{b}_j \lambda_{\star,j} + \mathfrak{b}_M \lambda_M + \sum_{h=1}^{l} \mathfrak{c}_h \underline{\mathfrak{t}}_{\star,h} \right) w(x,y,t) = 0.
\end{aligned}$$

It follows that $w \in \mathcal{H}$. Thus, in light of (5.12) we have

$$0 = \theta \cdot \partial^K w(0) = \sum_{|\iota| \le K} \theta_\iota \partial^\iota w(0) = \sum_{|i|+|I|+|\mathfrak{I}| \le K} \theta_{i,I,\mathfrak{I}} \partial_x^i \partial_y^I \partial_t^{\mathfrak{I}} w(0). \tag{5.49}$$

Now, we recall (5.40) and we claim that, for any $j \in \{1, \dots, n\}$, any $\ell \in \{1, \dots, p_j\}$, and any $i_{j\ell} \in \mathbb{N}$, we have

$$\partial_{x_{j\ell}}^{i_{j\ell}} \overline{v}_{j\ell}(0) \ne 0. \tag{5.50}$$

We prove it by induction over $i_{j\ell}$. Indeed, if $i_{j\ell} \in \{0, \dots, r_{j\ell} - 1\}$, then the initial condition in (5.38) (if $\ell = 1$) or (5.39) (if $\ell \ge 2$) gives $\partial_{x_{i\ell}}^{i_{j\ell}} \overline{v}_{i\ell}(0) = 1$, and so (5.50) is true in this case.

To perform the inductive step, let us now suppose that the claim in (5.50) still holds for all $i_{j\ell} \in \{0, \dots, i_0\}$ for some i_0 such that $i_0 \ge r_{j\ell} - 1$. Then, using the equation in (5.38) (if $\ell = 1$) or in (5.39) (if $\ell \ge 2$), we have

$$\partial_{x_{j\ell}}^{i_0+1} \overline{v}_j = \partial_{x_{j\ell}}^{i_0+1-r_{j\ell}} \partial_{x_{j\ell}}^{r_{j\ell}} \overline{v}_j = -\tilde{a}_j \partial_{x_{j\ell}}^{i_0+1-r_{j\ell}} \overline{v}_j, \tag{5.51}$$

with

$$\tilde{a}_j := \begin{cases} \overline{a}_j & \text{if } \ell = 1, \\ -1 & \text{if } \ell \ge 2. \end{cases}$$

Note that $\tilde{a}_j \ne 0$, in view of (5.36), and $\partial_{x_{j\ell}}^{i_0+1-r_{j\ell}} \overline{v}_j(0) \ne 0$, by the inductive assumption. These considerations and (5.51) give $\partial_{x_{j\ell}}^{i_0+1} \overline{v}_j(0) \ne 0$, and this proves (5.50).

Now, using (5.40) and (5.50) we have, for any $j \in \{1, \dots, n\}$ and any $i_j \in \mathbb{N}^{p_j}$,

$$\partial_{x_j}^{i_j} \overline{v}_j(0) \ne 0.$$

From this, (5.20), and the computation in (5.44) it follows that, for any $j \in \{1, \dots, n\}$ and any $i_j \in \mathbb{N}^{p_j}$,

$$\partial_{x_j}^{i_j} v_j(0) = \underline{\mathcal{X}}_j^{i_j} \partial_{x_j}^{i_j} \overline{v}_j(0) \ne 0. \tag{5.52}$$

We also note that, in light of (5.33), (5.47), and (5.49),

$$\begin{aligned} 0 = & \sum_{|i|+|I|+|\mathfrak{I}| \le K} \theta_{i,I,\mathfrak{I}} \partial_{x_1}^{i_1} v_1(0) \cdots \partial_{x_n}^{i_n} v_n(0) \partial_{y_1}^{I_1} \phi_1(e_1 + \epsilon Y_1) \cdots \partial_{y_M}^{I_M} \phi_M(e_M + \epsilon Y_M) \\ & \times \partial_{t_1}^{\mathfrak{I}_1} \psi_1(0) \cdots \partial^{\mathfrak{I}_l} \psi_l(0). \end{aligned} \tag{5.53}$$

Now, by (5.27) and Proposition 4.10 (applied to $s := s_j$, $\beta := I_j$, $e := \frac{e_j}{\omega_j} \in \partial B_1^{m_j}$, due to (5.45), and $X := \frac{Y_j}{\omega_j}$), we see that, for any $j = 1, \dots, M$,

$$\begin{aligned} \omega_j^{|I_j|} \lim_{\epsilon \searrow 0} \epsilon^{|I_j| - s_j} \partial_{y_j}^{I_j} \phi_j(e_j + \epsilon Y_j) &= \lim_{\epsilon \searrow 0} \epsilon^{|I_j| - s_j} \partial_{y_j}^{I_j} \tilde{\phi}_{\star,j}\left(\frac{e_j + \epsilon Y_j}{\omega_j} \right) \\ &= \kappa_j \frac{e_j^{I_j}}{\omega_j^{|I_j|}} \left(-\frac{e_j}{\omega_j} \cdot \frac{Y_j}{\omega_j} \right)_+^{s_j - |I_j|}, \end{aligned} \tag{5.54}$$

with $\kappa_j \ne 0$, in the sense of distributions (in the coordinates Y_j).

Moreover, using (5.35) and (5.48), it follows that

$$\begin{aligned}\partial_{t_h}^{\mathfrak{I}_h}\psi_h(0) &= \sum_{j=0}^{+\infty} \frac{\underline{t}_{\star,h}^{j}\alpha_h j(\alpha_h j-1)\dots(\alpha_h j-\mathfrak{I}_h+1)(0-a_h)^{\alpha_h j-\mathfrak{I}_h}}{\Gamma(\alpha_h j+1)} \\ &= \sum_{j=0}^{+\infty} \frac{\underline{t}_{\star,h}^{j}\alpha_h j(\alpha_h j-1)\dots(\alpha_h j-\mathfrak{I}_h+1)\epsilon^{\alpha_h j-\mathfrak{I}_h}}{\Gamma(\alpha_h j+1)\underline{t}_h^{\alpha_h j-\mathfrak{I}_h}} \\ &= \sum_{j=1}^{+\infty} \frac{\underline{t}_{\star,h}^{j}\alpha_h j(\alpha_h j-1)\dots(\alpha_h j-\mathfrak{I}_h+1)\epsilon^{\alpha_h j-\mathfrak{I}_h}}{\Gamma(\alpha_h j+1)\underline{t}_h^{\alpha_h j-\mathfrak{I}_h}}.\end{aligned}$$

Accordingly, recalling (5.30), we obtain

$$\begin{aligned}\lim_{\epsilon\searrow 0}\epsilon^{\mathfrak{I}_h-\alpha_h}\partial_{t_h}^{\mathfrak{I}_h}\psi_h(0) &= \lim_{\epsilon\searrow 0}\sum_{j=1}^{+\infty} \frac{\underline{t}_{\star,h}^{j}\alpha_h j(\alpha_h j-1)\dots(\alpha_h j-\mathfrak{I}_h+1)\epsilon^{\alpha_h(j-1)}}{\Gamma(\alpha_h j+1)\underline{t}_h^{\alpha_h j-\mathfrak{I}_h}} \\ &= \frac{\underline{t}_{\star,h}\alpha_h(\alpha_h-1)\dots(\alpha_h-\mathfrak{I}_h+1)}{\Gamma(\alpha_h+1)\underline{t}_h^{\alpha_h-\mathfrak{I}_h}} = \frac{\underline{t}_h^{\mathfrak{I}_h}\alpha_h(\alpha_h-1)\dots(\alpha_h-\mathfrak{I}_h+1)}{\Gamma(\alpha_h+1)}.\end{aligned} \tag{5.55}$$

Also, recalling (5.13), we can write (5.53) as

$$\begin{aligned}0 = &\sum_{\substack{|i|+|I|+|\mathfrak{I}|\le K\\ |I|\le|\bar{I}|}} \theta_{i,I,\mathfrak{I}}\partial_{x_1}^{i_1}v_1(0)\cdots\partial_{x_n}^{i_n}v_n(0)\partial_{y_1}^{I_1}\phi_1(e_1+\epsilon Y_1)\cdots\partial_{y_M}^{I_M}\phi_M(e_M+\epsilon Y_M) \\ &\times\partial_{t_1}^{\mathfrak{I}_1}\psi_1(0)\cdots\partial_{t_l}^{\mathfrak{I}_l}\psi_l(0).\end{aligned} \tag{5.56}$$

Moreover, we define

$$\Xi := |\bar{I}| - \sum_{j=1}^{M} s_j + |\mathfrak{I}| - \sum_{h=1}^{l}\alpha_h.$$

Then, we multiply (5.56) by $\epsilon^{\Xi}\in(0,+\infty)$, and we send ϵ to zero. In this way, from (5.54), (5.55), and (5.56) we obtain

$$\begin{aligned}0 = &\lim_{\epsilon\searrow 0}\epsilon^{\Xi}\sum_{\substack{|i|+|I|+|\mathfrak{I}|\le K\\ |I|\le|\bar{I}|}} \theta_{i,I,\mathfrak{I}}\partial_{x_1}^{i_1}v_1(0)\cdots\partial_{x_n}^{i_n}v_n(0)\partial_{y_1}^{I_1}\phi_1(e_1+\epsilon Y_1)\cdots\partial_{y_M}^{I_M}\phi_M(e_M+\epsilon Y_M) \\ &\times\partial_{t_1}^{\mathfrak{I}_1}\psi_1(0)\cdots\partial_{t_l}^{\mathfrak{I}_l}\psi_l(0) \\ = &\lim_{\epsilon\searrow 0}\sum_{\substack{|i|+|I|+|\mathfrak{I}|\le K\\ |I|\le|\bar{I}|}} \epsilon^{|\bar{I}|-|I|}\theta_{i,I,\mathfrak{I}}\partial_{x_1}^{i_1}v_1(0)\cdots\partial_{x_n}^{i_n}v_n(0) \\ &\times\epsilon^{|I_1|-s_1}\partial_{y_1}^{I_1}\phi_1(e_1+\epsilon Y_1)\cdots\epsilon^{|I_M|-s_M}\partial_{y_M}^{I_M}\phi_M(e_M+\epsilon Y_M) \\ &\times\epsilon^{\mathfrak{I}_1-\alpha_1}\partial_{t_1}^{\mathfrak{I}_1}\psi_1(0)\cdots\epsilon^{\mathfrak{I}_l-\alpha_l}\partial_{t_l}^{\mathfrak{I}_l}\psi_l(0)\end{aligned}$$

$$
\begin{aligned}
&= \sum_{\substack{|i|+|I|+|\mathfrak{J}|\le K\\ |I|=|\bar I|}} \tilde C_{i,I,\mathfrak{J}}\theta_{i,I,\mathfrak{J}}\partial^{i_1}_{x_1}v_1(0)\cdots\partial^{i_n}_{x_n}v_n(0)\\
&\quad\times e_1^{I_1}\cdots e_M^{I_M}(-e_1\cdot Y_1)_+^{s_1-|I_1|}\cdots(-e_M\cdot Y_M)_+^{s_M-|I_M|}\underline{t}_1^{\mathfrak{J}_1}\cdots\underline{t}_l^{\mathfrak{J}_l},
\end{aligned}
$$

for a suitable $\tilde C_{i,I,\mathfrak{J}}\neq 0$ (strictly speaking, the above identity holds in the sense of distribution with respect to the coordinates Y and $\underline{t}$, but since the left hand side vanishes, we can consider it also a pointwise identity).

Hence, recalling (5.52),

$$
\begin{aligned}
0 &= \sum_{\substack{|i|+|I|+|\mathfrak{J}|\le K\\ |I|=|\bar I|}} C_{i,I,\mathfrak{J}}\theta_{i_1,\dots,i_n,I_1,\dots,I_M,\mathfrak{J}_1,\dots,\mathfrak{J}_l}\underline{x}_1^{i_1}\cdots\underline{x}_n^{i_n}\\
&\quad\times e_1^{I_1}\cdots e_M^{I_M}(-e_1\cdot Y_1)_+^{s_1-|I_1|}\cdots(-e_M\cdot Y_M)_+^{s_M-|I_M|}\underline{t}_1^{\mathfrak{J}_1}\cdots\underline{t}_l^{\mathfrak{J}_l}\\
&= (-e_1\cdot Y_1)_+^{s_1}\cdots(-e_M\cdot Y_M)_+^{s_M}\\
&\quad\times \sum_{\substack{|i|+|I|+|\mathfrak{J}|\le K\\ |I|=|\bar I|}} C_{i,I,\mathfrak{J}}\theta_{i,I,\mathfrak{J}}\underline{x}^i e^I(-e_1\cdot Y_1)_+^{-|I_1|}\cdots(-e_M\cdot Y_M)_+^{-|I_M|}\underline{t}^{\mathfrak{J}},
\end{aligned}
\tag{5.57}
$$

for a suitable $C_{i,I,\mathfrak{J}}\neq 0$.

We observe that the equality in (5.57) is valid for any choice of the free parameters $(\underline{x},Y,\underline{t})$ in an open subset of $\mathbb{R}^{p_1+\dots+p_n}\times\mathbb{R}^{m_1+\dots+m_M}\times\mathbb{R}^l$, as prescribed in (5.20), (5.21), and (5.46).

Now, we take new free parameters, $\underline{y}_1,\dots,\underline{y}_M$ with $\underline{y}_j\in\mathbb{R}^{m_j}\setminus\{0\}$, and we define

$$
e_j := \frac{\omega_j\underline{y}_j}{|\underline{y}_j|}\quad\text{and}\quad Y_j := -\frac{\underline{y}_j}{|\underline{y}_j|^2}. \tag{5.58}
$$

We stress that the setting in (5.58) is compatible with that in (5.46), since

$$
e_j\cdot Y_j = -\frac{\omega_j\underline{y}_j}{|\underline{y}_j|}\cdot\frac{\underline{y}_j}{|\underline{y}_j|^2} = -\frac{\omega_j}{|\underline{y}_j|} < 0,
$$

thanks to (5.26). We also note that, for all $j\in\{1,\dots,M\}$,

$$
e_j^{I_j}(-e_j\cdot Y_j)_+^{-|I_j|} = \frac{\omega_j^{|I_j|}\underline{y}_j^{I_j}}{|\underline{y}_j|^{|I_j|}}\,\frac{|\underline{y}_j|^{|I_j|}}{\omega_j^{|I_j|}} = \underline{y}_j^{I_j},
$$

and hence

$$
e^I(-e_1\cdot Y_1)_+^{-|I_1|}\cdots(-e_M\cdot Y_M)_+^{-|I_M|} = \underline{y}^I.
$$

Plugging this into formula (5.57), we obtain the first identity in (5.15), as desired. Hence, the proof of (5.15) in case 1 is complete. □

Proof of (5.15), *case 2.* Thanks to the assumptions given in case 2, we can suppose that formula (5.17) still holds, and also that

$$c_l > 0. \tag{5.59}$$

In addition, for any $j \in \{1, \ldots, M\}$, we consider λ_j and ϕ_j as in (5.28).

Then, we define

$$R := \left(\frac{1}{|a_1|} \left(\sum_{h=1}^{l-1} |c_h| + \sum_{j=1}^{M} |b_j| \lambda_j \right) \right)^{1/|r_1|}. \tag{5.60}$$

We note that, in light of (5.17), the setting in (5.60) is well-defined.

Now, we fix two sets of free parameters $\underline{x}_1, \ldots, \underline{x}_n$ as in (5.20) and $\underline{t}_{\star,1}, \ldots, \underline{t}_{\star,l}$ as in (5.21), here taken with R as in (5.60). Moreover, we define

$$\lambda := \frac{1}{c_l \underline{t}_{\star,l}} \left(\sum_{j=1}^{n} |a_j| \underline{x}_j^{r_j} - \sum_{j=1}^{M} b_j \lambda_j - \sum_{h=1}^{l-1} c_h \underline{t}_{\star,h} \right). \tag{5.61}$$

We note that (5.61) is well-defined, thanks to (5.21) and (5.59). Furthermore, recalling (5.20), (5.24), and (5.60), we obtain

$$\begin{aligned} \sum_{i=1}^{n} |a_i| \underline{x}_i^{r_i} &\geq |a_1| \underline{x}_1^{r_1} > |a_1| (R+1)^{|r_1|} > |a_1| R^{|r_1|} \\ &= \sum_{h=1}^{l-1} |c_h| + \sum_{j=1}^{M} |b_j| \lambda_j \geq \sum_{h=1}^{l-1} c_h \underline{t}_{\star,h} + \sum_{j=1}^{M} b_j \lambda_j. \end{aligned}$$

Consequently, by (5.61),

$$\lambda > 0. \tag{5.62}$$

Hence, we can define

$$\overline{\lambda} := \lambda^{1/\alpha_l}. \tag{5.63}$$

Moreover, we consider $a_h \in (-2, 0)$, for every $h \in \{1, \ldots, l\}$, to be chosen appropriately in what follows (the exact choice will be performed in (5.70)), and, using the notation in (5.29) and (5.30), we define

$$\psi_h(t_h) := \psi_{\star,h}(\underline{t}_h(t_h - a_h)) = E_{\alpha_h,1}(\underline{t}_{\star,h}(t_h - a_h)^{\alpha_h}) \quad \text{if } h \in \{1, \ldots, l-1\} \tag{5.64}$$

and

$$\psi_l(t_l) := \psi_{\star,l}(\overline{\lambda} \underline{t}_l(t_l - a_l)) = E_{\alpha_l,1}(\lambda \underline{t}_{\star,l}(t_l - a_l)^{\alpha_l}). \tag{5.65}$$

We recall that, thanks to Lemma 3.2, the function in (5.64) solves (5.32) and satisfies (5.35) for any $h \in \{1, \dots, l-1\}$, while the function in (5.65) solves

$$\begin{cases} D^{\alpha_l}_{t_l,a_l} \psi_l(t_l) = \lambda \mathfrak{t}_{\star,l} \psi_l(t_l) & \text{in } (a_l, +\infty), \\ \psi_l(a_l) = 1, \\ \partial^m_{t_l} \psi_l(a_l) = 0 & \text{for every } m \in \{1, \dots, [\alpha_l]\}. \end{cases} \tag{5.66}$$

As in (5.33), we extend the functions ψ_h constantly in $(-\infty, a_h)$, calling $\psi^\star_h$ this extended function. In this way, Lemma A.3 in [20] translates (5.66) into

$$D^{\alpha_h}_{t_h,-\infty} \psi^\star_h(t_h) = \mathfrak{t}_{\star,h} \psi_h(t_h) = \mathfrak{t}_{\star,h} \psi^\star_h(t_h) \quad \text{in every interval } I \Subset (a_h, +\infty). \tag{5.67}$$

Now, we let $\epsilon > 0$, to be taken small possibly depending on the free parameters, and we exploit the functions defined in (5.42) and (5.43), provided that one replaces the positive constant R defined in (5.19) with the one in (5.60), when necessary.

With this idea in mind, for any $j \in \{1, \dots, M\}$, we let[2]

$$e_j \in \partial B^{m_j}_1, \tag{5.68}$$

and we define

$$\begin{aligned} w(x,y,t) := {} & \tau_1(x) v_1(x_1) \cdots v_n(x_n) \phi_1(y_1 + e_1 + \epsilon Y_1) \cdots \phi_M(y_M + e_M + \epsilon Y_M) \\ & \times \psi^\star_1(t_1) \cdots \psi^\star_l(t_l), \end{aligned} \tag{5.69}$$

where the setting in (5.28), (5.42), (5.43), (5.46), (5.64), and (5.65) has been exploited.

We also note that $w \in C(\mathbb{R}^N) \cap C_0(\mathbb{R}^{N-l}) \cap \mathcal{A}$. Moreover, if

$$a = (a_1, \dots, a_l) := \left(-\frac{\epsilon}{\mathfrak{t}_1}, \dots, -\frac{\epsilon}{\mathfrak{t}_l} \right) \in (-\infty, 0)^l, \tag{5.70}$$

(x,y) is sufficiently close to the origin, and $t \in (a_1, +\infty) \times \cdots \times (a_l, +\infty)$, we have

$$\begin{aligned} & \Lambda_{-\infty} w(x,y,t) \\ & = \left(\sum_{i=1}^n \mathfrak{a}_i \partial^{r_i}_{x_i} + \sum_{j=1}^M \mathfrak{b}_j (-\Delta)^{s_j}_{y_j} + \sum_{h=1}^l \mathfrak{c}_h D^{\alpha_h}_{t_h,-\infty} \right) w(x,y,t) \\ & = \sum_{i=1}^n \mathfrak{a}_i v_1(x_1) \cdots v_{i-1}(x_{i-1}) \partial^{r_i}_{x_i} v_i(x_i) v_{i+1}(x_{i+1}) \cdots v_n(x_n) \\ & \quad \times \phi_1(y_1 + e_1 + \epsilon Y_1) \cdots \phi_M(y_M + e_M + \epsilon Y_M) \psi^\star_1(t_1) \cdots \psi^\star_{l-1}(t_{l-1}) \psi^\star_l(t_l) \\ & \quad + \sum_{j=1}^M \mathfrak{b}_j v_1(x_1) \cdots v_n(x_n) \phi_1(y_1 + e_1 + \epsilon Y_1) \cdots \phi_{j-1}(y_{j-1} + e_{j-1} + \epsilon Y_{j-1}) \end{aligned}$$

2 Comparing (5.68) with (5.45), we observe that (5.45) reduces to (5.68) with the choice $\omega_j := 1$.

$$
\begin{aligned}
&\times (-\Delta)^{s_j}_{y_j}\phi_j(y_j+e_j+\epsilon Y_j)\phi_{j+1}(y_{j+1}+e_{j+1}+\epsilon Y_{j+1})\cdots\phi_M(y_M+e_M+\epsilon Y_M)\\
&\times \psi_1^\star(t_1)\cdots\psi_{l-1}^\star(t_{l-1})\psi_l^\star(t_l)\\
&+\sum_{h=1}^{l}\mathscr{c}_h v_1(x_1)\cdots v_n(x_n)\phi_1(y_1+e_1+\epsilon Y_1)\cdots\phi_M(y_M+e_M+\epsilon Y_M)\psi_1^\star(t_1)\cdots\psi_{h-1}^\star(t_{h-1})\\
&\times D^{\alpha_h}_{t_h,-\infty}\psi_h^\star(t_h)\psi_{h+1}^\star(t_{h+1})\cdots\psi_{l-1}^\star(t_{l-1})\psi_l^\star(t_l)\\
=&-\sum_{i=1}^{n}\mathscr{a}_i\overline{\mathscr{a}}_i\underline{\mathscr{x}}_i^{r_i}v_1(x_1)\cdots v_n(x_n)\phi_1(y_1+e_1+\epsilon Y_1)\cdots\phi_M(y_M+e_M+\epsilon Y_M)\\
&\times \psi_1^\star(t_1)\cdots\psi_{l-1}^\star(t_{l-1})\psi_l^\star(t_l)\\
&+\sum_{j=1}^{M}\mathscr{b}_j\lambda_j v_1(x_1)\cdots v_n(x_n)\phi_1(y_1+e_1+\epsilon Y_1)\cdots\phi_M(y_M+e_M+\epsilon Y_M)\\
&\times \psi_1^\star(t_1)\cdots\psi_{l-1}^\star(t_{l-1})\psi_l^\star(t_l)\\
&+\sum_{h=1}^{l-1}\mathscr{c}_h\underline{\mathscr{t}}_{\star,h}v_1(x_1)\cdots v_n(x_n)\phi_1(y_1+e_1+\epsilon Y_1)\cdots\phi_M(y_M+e_M+\epsilon Y_M)\\
&\times \psi_1^\star(t_1)\cdots\psi_{l-1}^\star(t_{l-1})\psi_l^\star(t_l)\\
&+\mathscr{c}_l\lambda\underline{\mathscr{t}}_{\star,l}v_1(x_1)\cdots v_n(x_n)\phi_1(y_1+e_1+\epsilon Y_1)\cdots\phi_M(y_M+e_M+\epsilon Y_M)\\
&\times \psi_1^\star(t_1)\cdots\psi_{l-1}^\star(t_{l-1})\psi_l^\star(t_l)\\
=&\left(-\sum_{i=1}^{n}\mathscr{a}_i\overline{\mathscr{a}}_i\underline{\mathscr{x}}_i^{r_i}+\sum_{j=1}^{M}\mathscr{b}_j\lambda_j+\sum_{h=1}^{l-1}\mathscr{c}_h\underline{\mathscr{t}}_{\star,h}+\mathscr{c}_l\lambda\underline{\mathscr{t}}_{\star,l}\right)w(x,y,t),
\end{aligned}
$$

thanks to (5.28), (5.32), (5.44), and (5.67).

Consequently, making use of (5.37) and (5.61), when (x,y) is near the origin and $t\in(a_1,+\infty)\times\cdots\times(a_l,+\infty)$, we have

$$
\Lambda_{-\infty}w(x,y,t)=\left(-\sum_{i=1}^{n}|\mathscr{a}_i|\underline{\mathscr{x}}_i^{r_i}+\sum_{j=1}^{M}\mathscr{b}_j\lambda_j+\sum_{h=1}^{l-1}\mathscr{c}_h\underline{\mathscr{t}}_{\star,h}+\lambda\mathscr{c}_l\underline{\mathscr{t}}_{\star,l}\right)w(x,y,t)=0.
$$

It follows that $w\in\mathcal{H}$. Thus, in light of (5.12) we have

$$
0=\theta\cdot\partial^K w(0)=\sum_{|\iota|\le K}\theta_\iota\partial^\iota w(0)=\sum_{|i|+|I|+|\mathfrak{I}|\le K}\theta_{i,I,\mathfrak{I}}\partial^i_x\partial^I_y\partial^{\mathfrak{I}}_t w(0).
$$

Hence, in view of (5.52) and (5.69),

$$
\begin{aligned}
0=&\sum_{|i|+|I|+|\mathfrak{I}|\le K}\theta_{i,I,\mathfrak{I}}\partial^{i_1}_{x_1}v_1(0)\cdots\partial^{i_n}_{x_n}v_n(0)\\
&\times\partial^{I_1}_{y_1}\phi_1(e_1+\epsilon Y_1)\cdots\partial^{I_M}_{y_M}\phi_M(e_M+\epsilon Y_M)\partial^{\mathfrak{I}_1}_{t_1}\psi_1(0)\cdots\partial^{\mathfrak{I}_l}_{t_l}\psi_l(0)\\
=&\sum_{|i|+|I|+|\mathfrak{I}|\le K}\theta_{i,I,\mathfrak{I}}\underline{\mathscr{x}}_1^{r_1}\cdots\underline{\mathscr{x}}_n^{r_n}\partial^{i_1}_{x_1}\overline{v}_1(0)\cdots\partial^{i_n}_{x_n}\overline{v}_n(0)\\
&\times\partial^{I_1}_{y_1}\phi_1(e_1+\epsilon Y_1)\cdots\partial^{I_M}_{y_M}\phi_M(e_M+\epsilon Y_M)\partial^{\mathfrak{I}_1}_{t_1}\psi_1(0)\cdots\partial^{\mathfrak{I}_l}_{t_l}\psi_l(0).
\end{aligned}
\tag{5.71}
$$

Moreover, using (3.1), (5.65), and (5.70), it follows that

$$\begin{aligned}\partial_{t_l}^{\mathfrak{I}_l}\psi_l(0) &= \sum_{j=0}^{+\infty}\frac{\lambda^j \underline{t}_{\star,l}^j \alpha_l j(\alpha_l j-1)\dots(\alpha_l j-\mathfrak{I}_l+1)(0-a_l)^{\alpha_l j-\mathfrak{I}_l}}{\Gamma(\alpha_l j+1)}\\ &= \sum_{j=0}^{+\infty}\frac{\lambda^j \underline{t}_{\star,l}^j \alpha_l j(\alpha_l j-1)\dots(\alpha_l j-\mathfrak{I}_l+1)\epsilon^{\alpha_l j-\mathfrak{I}_l}}{\Gamma(\alpha_l j+1)\underline{t}_l^{\alpha_l j-\mathfrak{I}_l}}\\ &= \sum_{j=1}^{+\infty}\frac{\lambda^j \underline{t}_{\star,l}^j \alpha_l j(\alpha_l j-1)\dots(\alpha_l j-\mathfrak{I}_l+1)\epsilon^{\alpha_l j-\mathfrak{I}_l}}{\Gamma(\alpha_l j+1)\underline{t}_l^{\alpha_l j-\mathfrak{I}_l}}.\end{aligned}$$

Accordingly, by (5.30), we obtain

$$\begin{aligned}\lim_{\epsilon\searrow 0}\epsilon^{\mathfrak{I}_l-\alpha_l}\partial_{t_l}^{\mathfrak{I}_l}\psi_l(0) &= \lim_{\epsilon\searrow 0}\sum_{j=1}^{+\infty}\frac{\lambda^j \underline{t}_{\star,l}^j \alpha_l j(\alpha_l j-1)\dots(\alpha_l j-\mathfrak{I}_l+1)\epsilon^{\alpha_l(j-1)}}{\Gamma(\alpha_l j+1)\underline{t}_l^{\alpha_l j-\mathfrak{I}_l}}\\ &= \frac{\lambda \underline{t}_{\star,l}\alpha_l(\alpha_l-1)\dots(\alpha_l-\mathfrak{I}_l+1)}{\Gamma(\alpha_l+1)\underline{t}_l^{\alpha_l-\mathfrak{I}_l}} = \frac{\lambda \underline{t}_l^{\mathfrak{I}_l}\alpha_l(\alpha_l-1)\dots(\alpha_l-\mathfrak{I}_l+1)}{\Gamma(\alpha_l+1)}.\end{aligned} \tag{5.72}$$

Hence, recalling (5.14), we can write (5.71) as

$$\begin{aligned}0 = &\sum_{\substack{|i|+|I|+|\mathfrak{I}|\le K\\ |\mathfrak{I}|\le|\overline{\mathfrak{I}}|}}\theta_{i,I,\mathfrak{I}}\,\underline{x}_1^{r_1}\cdots\underline{x}_n^{r_n}\partial_{x_1}^{i_1}\overline{v}_1(0)\cdots\partial_{x_n}^{i_n}\overline{v}_n(0)\\ &\times\partial_{y_1}^{I_1}\phi_1(e_1+\epsilon Y_1)\cdots\partial_{y_M}^{I_M}\phi_M(e_M+\epsilon Y_M)\partial_{t_1}^{\mathfrak{I}_1}\psi_1(0)\cdots\partial_{t_l}^{\mathfrak{I}_l}\psi_l(0).\end{aligned} \tag{5.73}$$

Moreover, we define

$$\Xi := |\overline{\mathfrak{I}}| - \sum_{h=1}^{l}\alpha_h + |I| - \sum_{j=1}^{M}s_j.$$

Then, we multiply (5.73) by $\epsilon^{\Xi}\in(0,+\infty)$, and we send ϵ to zero. In this way, from (5.55), used here for $h\in\{1,\dots,l-1\}$, (5.72), and (5.73), we obtain

$$\begin{aligned}0 = &\lim_{\epsilon\searrow 0}\epsilon^{\Xi}\sum_{\substack{|i|+|I|+|\mathfrak{I}|\le K\\ |\mathfrak{I}|\le|\overline{\mathfrak{I}}|}}\theta_{i,I,\mathfrak{I}}\,\underline{x}_1^{r_1}\cdots\underline{x}_n^{r_n}\partial_{x_1}^{i_1}\overline{v}_1(0)\cdots\partial_{x_n}^{i_n}\overline{v}_n(0)\\ &\times\partial_{y_1}^{I_1}\phi_1(e_1+\epsilon Y_1)\cdots\partial_{y_M}^{I_M}\phi_M(e_M+\epsilon Y_M)\\ &\times\partial_{t_1}^{\mathfrak{I}_1}\psi_1(0)\cdots\partial_{t_l}^{\mathfrak{I}_l}\psi_l(0)\\ = &\lim_{\epsilon\searrow 0}\sum_{\substack{|i|+|I|+|\mathfrak{I}|\le K\\ |\mathfrak{I}|\le|\overline{\mathfrak{I}}|}}\epsilon^{|\overline{\mathfrak{I}}|-|\mathfrak{I}|}\theta_{i,I,\mathfrak{I}}\,\underline{x}_1^{r_1}\cdots\underline{x}_n^{r_n}\partial_{x_1}^{i_1}\overline{v}_1(0)\cdots\partial_{x_n}^{i_n}\overline{v}_n(0)\\ &\times\epsilon^{|I_1|-s_1}\partial_{y_1}^{I_1}\phi_1(e_1+\epsilon Y_1)\cdots\epsilon^{|I_M|-s_M}\partial_{y_M}^{I_M}\phi_M(e_M+\epsilon Y_M)\end{aligned}$$

$$\begin{aligned}
&\times \epsilon^{\mathfrak{I}_1-\alpha_1}\partial_{t_1}^{\overline{\mathfrak{I}}_1}\psi_1(0)\cdots\epsilon^{\mathfrak{I}_l-\alpha_l}\partial_{t_l}^{\overline{\mathfrak{I}}_l}\psi_l(0)\\
&= \sum_{\substack{|i|+|I|+|\mathfrak{I}|\le K\\ |\mathfrak{I}|=|\overline{\mathfrak{I}}|}} \lambda\tilde{C}_{i,I,\mathfrak{I}}\theta_{i,I,\mathfrak{I}}\,\underline{x}_1^{r_1}\cdots\underline{x}_n^{r_n}\partial_{x_1}^{i_1}\overline{v}_1(0)\cdots\partial_{x_n}^{i_n}\overline{v}_n(0)\\
&\quad\times e_1^{I_1}\cdots e_M^{I_M}(-e_1\cdot Y_1)_+^{s_1-|I_1|}\cdots(-e_M\cdot Y_M)_+^{s_M-|I_M|}\underline{t}_1^{\mathfrak{I}_1}\cdots\underline{t}_l^{\mathfrak{I}_l},
\end{aligned}$$

for a suitable $\tilde{C}_{i,I,\mathfrak{I}}$. We stress that $\tilde{C}_{i,I,\mathfrak{I}} \neq 0$, thanks also to (5.54), applied here with $\omega_j := 1$, $\tilde{\phi}_{\star,j} := \phi_j$, and e_j as in (5.68) for any $j \in \{1,\dots,M\}$.

Hence, recalling (5.62),

$$\begin{aligned}
0 &= \sum_{\substack{|i|+|I|+|\mathfrak{I}|\le K\\ |\mathfrak{I}|=|\overline{\mathfrak{I}}|}} C_{i,I,\mathfrak{I}}\theta_{i_1,\dots,i_n,I_1,\dots,I_M,\mathfrak{I}_1,\dots,\mathfrak{I}_l}\,\underline{x}_1^{i_1}\cdots\underline{x}_n^{i_n}\\
&\quad\times e_1^{I_1}\cdots e_M^{I_M}(-e_1\cdot Y_1)_+^{s_1-|I_1|}\cdots(-e_M\cdot Y_M)_+^{s_M-|I_M|}\underline{t}_1^{\mathfrak{I}_1}\cdots\underline{t}_l^{\mathfrak{I}_l}\\
&= (-e_1\cdot Y_1)_+^{s_1}\cdots(-e_M\cdot Y_M)_+^{s_M}\\
&\quad\times \sum_{\substack{|i|+|I|+|\mathfrak{I}|\le K\\ |\mathfrak{I}|=|\overline{\mathfrak{I}}|}} C_{i,I,\mathfrak{I}}\theta_{i,I,\mathfrak{I}}\,\underline{x}^{i}e^{I}(-e_1\cdot Y_1)_+^{-|I_1|}\cdots(-e_M\cdot Y_M)_+^{-|I_M|}\underline{t}^{\mathfrak{I}},
\end{aligned}\tag{5.74}$$

for a suitable $C_{i,I,\mathfrak{I}} \neq 0$.

We observe that the equality in (5.74) is valid for any choice of the free parameters $(\underline{x}, Y, \underline{t})$ in an open subset of $\mathbb{R}^{p_1+\dots+p_n}\times\mathbb{R}^{m_1+\dots+m_M}\times\mathbb{R}^l$, as prescribed in (5.20), (5.21), and (5.46).

Now, we take new free parameters $\underline{y}_j$ with $\underline{y}_j \in \mathbb{R}^{m_j}\setminus\{0\}$ for any $j = 1,\dots,M$, and we perform in (5.74) the same change of variables as done in (5.58), obtaining

$$0 = \sum_{\substack{|i|+|I|+|\mathfrak{I}|\le K\\ |\mathfrak{I}|=|\overline{\mathfrak{I}}|}} C_{i,I,\mathfrak{I}}\theta_{i,I,\mathfrak{I}}\,\underline{x}^{i}\underline{y}^{I}\underline{t}^{\mathfrak{I}},$$

for some $C_{i,I,\mathfrak{I}} \neq 0$.

Hence, the second identity in (5.15) is obtained as desired, and the proof of Lemma 5.3 in case 2 is completed. □

Proof of (5.15), *case 3.* We divide the proof of case 3 into two subcases, namely, either

$$\text{there exists } h \in \{1,\dots,l\} \quad \text{such that } c_h \neq 0, \tag{5.75}$$

or

$$c_h = 0 \quad \text{for every } h \in \{1,\dots,l\}. \tag{5.76}$$

We start by dealing with the case in (5.75). Up to relabeling and reordering the coefficients c_h, we assume that

$$c_1 \neq 0. \tag{5.77}$$

Also, thanks to the assumptions given in case 3, we suppose that

$$b_M < 0, \tag{5.78}$$

and, for any $j \in \{1, \dots, M\}$, we consider $\lambda_{\star,j}$ and $\tilde{\phi}_{\star,j}$ as in (5.16). Then, we take $\omega_j := 1$ and ϕ_j as in (5.27), so that (5.28) is satisfied. In particular, here we have

$$\lambda_j = \lambda_{\star,j} \quad \text{and} \quad \phi_j = \tilde{\phi}_{\star,j}. \tag{5.79}$$

We define

$$R := \frac{1}{|c_1|} \sum_{j=1}^{M-1} |b_j| \lambda_{\star,j}. \tag{5.80}$$

We note that, in light of (5.77), the setting in (5.80) is well-defined.

Now, we fix a set of free parameters

$$\underline{t}_{\star,1} \in (R+1, R+2), \dots \underline{t}_{\star,l} \in (R+1, R+2). \tag{5.81}$$

Moreover, we define

$$\lambda_M := \frac{1}{b_M} \left(- \sum_{j=1}^{M-1} b_j \lambda_{\star,j} - \sum_{h=1}^{l} |c_h| \underline{t}_{\star,h} \right). \tag{5.82}$$

We note that (5.82) is well-defined thanks to (5.78). From (5.80) we deduce that

$$\begin{aligned} \sum_{h=1}^{l} |c_h| \underline{t}_{\star,h} + \sum_{j=1}^{M-1} b_j \lambda_{\star,j} &\geq |c_1| \underline{t}_{\star,1} - \sum_{j=1}^{M-1} |b_j| \lambda_{\star,j} \\ &> |c_1| R - \sum_{j=1}^{M-1} |b_j| \lambda_{\star,j} = 0. \end{aligned}$$

Consequently, by (5.78) and (5.82),

$$\lambda_M > 0. \tag{5.83}$$

Now, we define, for any $h \in \{1, \dots, l\}$,

$$\bar{c}_h := \begin{cases} \frac{c_h}{|c_h|} & \text{if } c_h \neq 0, \\ 1 & \text{if } c_h = 0. \end{cases}$$

We note that

$$\bar{c}_h \neq 0 \quad \text{for all } h \in \{1, \dots, l\}, \tag{5.84}$$

and

$$c_h \bar{c}_h = |c_h|. \tag{5.85}$$

Moreover, we consider $a_h \in (-2, 0)$, for every $h = 1, \dots, l$, to be chosen appropriately in what follows (see (5.93) for a precise choice).

Now, for every $h \in \{1, \dots, l\}$, we define

$$\psi_h(t_h) := E_{\alpha_h,1}(\bar{c}_h \underline{t}_{\star,h}(t_h - a_h)^{\alpha_h}), \tag{5.86}$$

where $E_{\alpha_h,1}$ denotes the Mittag-Leffler function with parameters $\alpha := \alpha_h$ and $\beta := 1$ as defined in (3.1). By Lemma 3.2, we know that

$$\begin{cases} D^{\alpha_h}_{t_h,a_h} \psi_h(t_h) = \bar{c}_h \underline{t}_{\star,h} \psi_h(t_h) & \text{in } (a_h, +\infty), \\ \psi_h(a_h) = 1, \\ \partial^m_{t_h} \psi_h(a_h) = 0 & \text{for any } m = 1, \dots, [\alpha_h], \end{cases} \tag{5.87}$$

and we consider again the extension $\psi_h^\star$ given in (5.33). By Lemma A.3 in [20], we know that (5.87) translates into

$$D^{\alpha_h}_{t_h,-\infty} \psi_h^\star(t_h) = \bar{c}_h \underline{t}_{\star,h} \psi_h^\star(t_h) \quad \text{in every interval } I \Subset (a_h, +\infty). \tag{5.88}$$

Now, we consider auxiliary parameters $\underline{t}_h$, e_j, and Y_j as in (5.30), (5.45), and (5.46). Moreover, we introduce an additional set of free parameters

$$\underline{\mathcal{X}} = (\underline{\mathcal{X}}_1, \dots, \underline{\mathcal{X}}_n) \in \mathbb{R}^{p_1} \times \dots \times \mathbb{R}^{p_n}. \tag{5.89}$$

We let $\epsilon > 0$, to be taken small possibly depending on the free parameters. We take $\tau \in C^\infty(\mathbb{R}^{p_1+\dots+p_n}, [0 + \infty))$ such that

$$\tau(x) := \begin{cases} \exp(\underline{\mathcal{X}} \cdot x) & \text{if } x \in B_1^{p_1+\dots+p_n}, \\ 0 & \text{if } x \in \mathbb{R}^{p_1+\dots+p_n} \setminus B_2^{p_1+\dots+p_n}, \end{cases} \tag{5.90}$$

where

$$\underline{\mathcal{X}} \cdot x := \sum_{j=1}^{n} \underline{\mathcal{X}}_i \cdot x_i$$

denotes the standard scalar product.

We note that, for any $i \in \mathbb{N}^{p_1} \times \dots \times \mathbb{N}^{p_n}$,

$$\partial^i_x \tau(0) = \partial^{i_1}_{x_1} \cdots \partial^{i_n}_{x_n} \tau(0) = \underline{\mathcal{X}}_{11}^{i_{11}} \cdots \underline{\mathcal{X}}_{1p_1}^{i_{1p_1}} \cdots \underline{\mathcal{X}}_{n1}^{i_{n1}} \cdots \underline{\mathcal{X}}_{np_n}^{i_{np_n}} = \underline{\mathcal{X}}^i. \tag{5.91}$$

We define

$$w(x, y, t) := \tau(x) \phi_1(y_1 + e_1 + \epsilon Y_1) \cdots \phi_M(y_M + e_M + \epsilon Y_M) \psi_1^\star(t_1) \cdots \psi_l^\star(t_l), \tag{5.92}$$

where the setting in (5.28) has also been exploited.

We also note that $w \in C(\mathbb{R}^N) \cap C_0(\mathbb{R}^{N-l}) \cap \mathcal{A}$. Moreover, if

$$a = (a_1, \dots, a_l) := \left(-\frac{\epsilon}{\underline{t}_1}, \dots, -\frac{\epsilon}{\underline{t}_l} \right) \in (-\infty, 0)^l, \tag{5.93}$$

(x, y) is sufficiently close to the origin, and $t \in (a_1, +\infty) \times \cdots \times (a_l, +\infty)$, we have

$$\begin{aligned}
&\Lambda_{-\infty} w(x,y,t) \\
&= \left(\sum_{j=1}^{M} b_j (-\Delta)_{y_j}^{s_j} + \sum_{h=1}^{l} c_h D_{t_h,-\infty}^{\alpha_h} \right) w(x,y,t) \\
&= \sum_{j=1}^{M} b_j \tau(x) \phi_1(y_1 + e_1 + \epsilon Y_1) \cdots \phi_{j-1}(y_{j-1} + e_{j-1} + \epsilon Y_{j-1}) (-\Delta)_{y_j}^{s_j} \phi_j(y_j + e_j + \epsilon Y_j) \\
&\quad \times \phi_{j+1}(y_{j+1} + e_{j+1} + \epsilon Y_{j+1}) \cdots \phi_M(y_M + e_M + \epsilon Y_M) \psi_1^\star(t_1) \cdots \psi_l^\star(t_l) \\
&\quad + \sum_{h=1}^{l} c_h \tau(x) \phi_1(y_1 + e_1 + \epsilon Y_1) \cdots \phi_M(y_M + e_M + \epsilon Y_M) \psi_1^\star(t_1) \cdots \psi_{h-1}^\star(t_{h-1}) \\
&\quad \times D_{t_h,-\infty}^{\alpha_h} \psi_h^\star(t_h) \psi_{h+1}^\star(t_{h+1}) \cdots \psi_l^\star(t_l) \\
&= \sum_{j=1}^{M} b_j \lambda_j \tau(x) \phi_1(y_1 + e_1 + \epsilon Y_1) \cdots \phi_M(y_M + e_M + \epsilon Y_M) \psi_1^\star(t_1) \cdots \psi_l^\star(t_l) \\
&\quad + \sum_{h=1}^{l} c_h \bar{c}_h \underline{t}_{\star,h} \tau(x) \phi_1(y_1 + e_1 + \epsilon Y_1) \cdots \phi_M(y_M + e_M + \epsilon Y_M) \psi_1^\star(t_1) \cdots \psi_l^\star(t_l) \\
&= \left(\sum_{j=1}^{M} b_j \lambda_j + \sum_{h=1}^{l} c_h \bar{c}_h \underline{t}_{\star,h} \right) w(x,y,t),
\end{aligned}$$

thanks to (5.28) and (5.88).

Consequently, making use of (5.79), (5.82), and (5.85), if (x, y) is near the origin and $t \in (a_1, +\infty) \times \cdots \times (a_l, +\infty)$, we have

$$\Lambda_{-\infty} w(x,y,t) = \left(\sum_{j=1}^{M} b_j \lambda_{\star,j} + b_M \lambda_M + \sum_{h=1}^{l} |c_h| \underline{t}_{\star,h} \right) w(x,y,t) = 0.$$

It follows that $w \in \mathcal{H}$. Thus, in light of (5.12) we have

$$0 = \theta \cdot \partial^K w(0) = \sum_{|\iota| \le K} \theta_\iota \partial^\iota w(0) = \sum_{|i|+|I|+|\mathfrak{I}| \le K} \theta_{i,I,\mathfrak{I}} \partial_x^i \partial_y^I \partial_t^{\mathfrak{I}} w(0).$$

From this and (5.92), we obtain

$$0 = \sum_{|i|+|I|+|\mathfrak{I}| \le K} \theta_{i,I,\mathfrak{I}} \partial_x^i \tau(0) \partial_{y_1}^{I_1} \phi_1(e_1 + \epsilon Y_1) \cdots \partial_{y_M}^{I_M} \phi_M(e_M + \epsilon Y_M) \partial_{t_1}^{\mathfrak{I}_1} \psi_1(0) \cdots \partial_{t_l}^{\mathfrak{I}_l} \psi_l(0). \tag{5.94}$$

Moreover, using (5.86) and (5.93), it follows that, for every $\mathfrak{I}_h \in \mathbb{N}$,

$$\begin{aligned}\partial_{t_h}^{\mathfrak{I}_h}\psi_h(0) &= \sum_{j=0}^{+\infty} \frac{\bar{c}_h^j \underline{t}_{\star,h}^j \alpha_h j(\alpha_h j-1)\dots(\alpha_h j-\mathfrak{I}_l+1)(0-a_h)^{\alpha_h j-\mathfrak{I}_h}}{\Gamma(\alpha_h j+1)} \\ &= \sum_{j=0}^{+\infty} \frac{\bar{c}_h^j \underline{t}_{\star,h}^j \alpha_h j(\alpha_h j-1)\dots(\alpha_h j-\mathfrak{I}_h+1)\epsilon^{\alpha_h j-\mathfrak{I}_h}}{\Gamma(\alpha_h j+1)\underline{t}_h^{\alpha_h j-\mathfrak{I}_h}} \\ &= \sum_{j=1}^{+\infty} \frac{\bar{c}_h^j \underline{t}_{\star,h}^j \alpha_h j(\alpha_h j-1)\dots(\alpha_h j-\mathfrak{I}_h+1)\epsilon^{\alpha_h j-\mathfrak{I}_h}}{\Gamma(\alpha_h j+1)\underline{t}_h^{\alpha_h j-\mathfrak{I}_h}}.\end{aligned}$$

Accordingly, recalling (5.30), we obtain

$$\begin{aligned}\lim_{\epsilon\searrow 0}\epsilon^{\mathfrak{I}_h-\alpha_h}\partial_{t_h}^{\mathfrak{I}_h}\psi_h(0) &= \lim_{\epsilon\searrow 0}\sum_{j=1}^{+\infty} \frac{\bar{c}_h^j \underline{t}_{\star,h}^j \alpha_h j(\alpha_h j-1)\dots(\alpha_h j-\mathfrak{I}_h+1)\epsilon^{\alpha_h (j-1)}}{\Gamma(\alpha_h j+1)\underline{t}_h^{\alpha_h j-\mathfrak{I}_h}} \\ &= \frac{\bar{c}_h \underline{t}_{\star,h}\alpha_h(\alpha_h-1)\dots(\alpha_h-\mathfrak{I}_h+1)}{\Gamma(\alpha_h+1)\underline{t}_h^{\alpha_h-\mathfrak{I}_h}} \\ &= \frac{\bar{c}_h \underline{t}_h^{\mathfrak{I}_h}\alpha_h(\alpha_h-1)\dots(\alpha_h-\mathfrak{I}_h+1)}{\Gamma(\alpha_h+1)}.\end{aligned} \tag{5.95}$$

Also, recalling (5.13), we can write (5.94) as

$$0 = \sum_{\substack{|i|+|I|+|\mathfrak{I}|\le K \\ |I|\le|\bar{I}|}} \theta_{i,I,\mathfrak{I}}\partial_x^i\tau(0)\partial_{y_1}^{I_1}\phi_1(e_1+\epsilon Y_1)\cdots\partial_{y_M}^{I_M}\phi_M(e_M+\epsilon Y_M)\partial_{t_1}^{\mathfrak{I}_1}\psi_1(0)\cdots\partial_{t_l}^{\mathfrak{I}_l}\psi_l(0). \tag{5.96}$$

Moreover, we define

$$\Xi := |\bar{I}| - \sum_{j=1}^{M} s_j + |\mathfrak{I}| - \sum_{h=1}^{l}\alpha_h.$$

Then, we multiply (5.96) by $\epsilon^{\Xi} \in (0,+\infty)$, and we send ϵ to zero. In this way, from (5.54), (5.91), (5.95), and (5.96) we obtain

$$\begin{aligned}0 &= \lim_{\epsilon\searrow 0}\epsilon^{\Xi}\sum_{\substack{|i|+|I|+|\mathfrak{I}|\le K \\ |I|\le|\bar{I}|}} \theta_{i,I,\mathfrak{I}}\partial_x^i\tau(0)\partial_{y_1}^{I_1}\phi_1(e_1+\epsilon Y_1)\cdots\partial_{y_M}^{I_M}\phi_M(e_M+\epsilon Y_M)\partial_{t_1}^{\mathfrak{I}_1}\psi_1(0)\cdots\partial_{t_l}^{\mathfrak{I}_l}\psi_l(0) \\ &= \lim_{\epsilon\searrow 0}\sum_{\substack{|i|+|I|+|\mathfrak{I}|\le K \\ |I|\le|\bar{I}|}} \epsilon^{|\bar{I}|-|I|}\theta_{i,I,\mathfrak{I}}\partial_x^i\tau(0)\epsilon^{|I_1|-s_1}\partial_{y_1}^{I_1}\phi_1(e_1+\epsilon Y_1)\cdots\epsilon^{|I_M|-s_M}\partial_{y_M}^{I_M}\phi_M(e_M+\epsilon Y_M) \\ &\quad\times\epsilon^{\mathfrak{I}_1-\alpha_1}\partial_{t_1}^{\mathfrak{I}_1}\psi_1(0)\cdots\epsilon^{\mathfrak{I}_l-\alpha_l}\partial_{t_l}^{\mathfrak{I}_l}\psi_l(0) \\ &= \sum_{\substack{|i|+|I|+|\mathfrak{I}|\le K \\ |I|=|\bar{I}|}} C_{i,I,\mathfrak{I}}\theta_{i,I,\mathfrak{I}}\,\underline{x}_1^{i_1}\cdots\underline{x}_n^{i_n}e_1^{I_1}\cdots e_M^{I_M}(-e_1\cdot Y_1)_+^{s_1-|I_1|}\cdots(-e_M\cdot Y_M)_+^{s_M-|I_M|}\underline{t}_1^{\mathfrak{I}_1}\cdots\underline{t}_l^{\mathfrak{I}_l}\end{aligned}$$

$$= (-e_1 \cdot Y_1)_+^{s_1} \cdots (-e_M \cdot Y_M)_+^{s_M}$$
$$\times \sum_{\substack{|i|+|I|+|\mathfrak{I}| \le K \\ |I|=|\bar{I}|}} C_{i,I,\mathfrak{I}} \theta_{i,I,\mathfrak{I}} \underline{x}^i e^I (-e_1 \cdot Y_1)_+^{-|I_1|} \cdots (-e_M \cdot Y_M)_+^{-|I_M|} \underline{t}^{\mathfrak{I}},$$

for a suitable $C_{i,I,\mathfrak{I}} \neq 0$.

We observe that the latter equality is valid for any choice of the free parameters $(\underline{x}, Y, \underline{t})$ in an open subset of $\mathbb{R}^{p_1+\cdots+p_n} \times \mathbb{R}^{m_1+\cdots+m_M} \times \mathbb{R}^l$, as prescribed in (5.46), (5.81), and (5.89).

Now, we take new free parameters $\underline{y}_j$ with $\underline{y}_j \in \mathbb{R}^{m_j} \setminus \{0\}$ for any $j = 1, \ldots, M$, and we perform in the latter identity the same change of variables as done in (5.58), obtaining

$$0 = \sum_{\substack{|i|+|I|+|\mathfrak{I}| \le K \\ |I|=|\bar{I}|}} C_{i,I,\mathfrak{I}} \theta_{i,I,\mathfrak{I}} \underline{x}^i \underline{y}^I \underline{t}^{\mathfrak{I}},$$

for some $C_{i,I,\mathfrak{I}} \neq 0$. This completes the proof of (5.15) in case (5.75) is satisfied.

Hence, we now focus on the case in which (5.76) holds. For any $j \in \{1, \ldots, M\}$, we consider the function $\psi \in H^{s_j}(\mathbb{R}^{m_j}) \cap C_0^{s_j}(\mathbb{R}^{m_j})$ constructed in Lemma 4.15 and we call such function ϕ_j, to make its dependence on j in explicit this case. We recall that

$$(-\Delta)_{y_j}^{s_j} \phi_j(y_j) = 0 \quad \text{in } B_1^{m_j}. \tag{5.97}$$

Also, for every $j \in \{1, \ldots, M\}$, we let e_j and Y_j be as in (5.45) and (5.46). Thanks to Lemma 4.15 and Remark 4.16, for any $I_j \in \mathbb{N}^{m_j}$, we know that

$$\lim_{\epsilon \searrow 0} \epsilon^{|I_j|-s_j} \partial_{y_j}^{I_j} \phi_j(e_j + \epsilon Y_j) = \kappa_{s_j} e_j^{I_j} (-e_j \cdot Y_j)_+^{s_j-|I_j|}, \tag{5.98}$$

for some $\kappa_{s_j} \neq 0$.

Moreover, for any $h = 1, \ldots, l$, we define $\overline{\tau}_h(t_h)$ as

$$\overline{\tau}_h(t_h) := \begin{cases} e^{\underline{t}_h t_h} & \text{if } t_h \in [-1, +\infty), \\ e^{-\underline{t}_h} \sum_{i=0}^{k_h-1} \frac{\underline{t}_h^i}{i!} (t_h + 1)^i & \text{if } t_h \in (-\infty, -1), \end{cases} \tag{5.99}$$

where $\underline{t} = (\underline{t}_1, \ldots, \underline{t}_l) \in (1, 2)^l$ are free parameters.

We note that, for any $h \in \{1, \ldots, l\}$ and $\mathfrak{I}_h \in \mathbb{N}$,

$$\partial_{t_h}^{\mathfrak{I}_h} \overline{\tau}_h(0) = \underline{t}_h^{\mathfrak{I}_h}. \tag{5.100}$$

Now, we define

$$w(x, y, t) := \tau(x) \phi_1(y_1 + e_1 + \epsilon Y_1) \cdots \phi_M(y_M + e_M + \epsilon Y_M) \overline{\tau}_1(t_1) \cdots \overline{\tau}_l(t_l), \tag{5.101}$$

where the setting of (5.27), (5.90), and (5.99) has been exploited. We have $w \in \mathcal{A}$. Moreover, we point out that, since τ, $\phi_1, \ldots, \phi_M$ are compactly supported, we have $w \in$

$C(\mathbb{R}^N) \cap C_0(\mathbb{R}^{N-l})$, and, using Proposition 4.18, for any $j \in \{1, \dots, M\}$, we have $\phi_j \in C^\infty(\mathcal{N}_j)$ for some neighborhood $\mathcal{N}_j$ of the origin in $\mathbb{R}^{m_j}$. Hence $w \in C^\infty(\mathcal{N})$.

Furthermore, using (5.97), when y is in a neighborhood of the origin we have

$$\begin{aligned}\Lambda_{-\infty} w(x,y,t) &= \tau(x)(\mathscr{b}_1(-\Delta)^{s_1}_{y_1}\phi_1(y_1+e_1+\epsilon Y_1))\cdots\phi_M(y_M+e_M+\epsilon Y_M)\overline{\tau}_1(t_1)\cdots\overline{\tau}_l(t_l)\\ &\quad+\dots+\tau(x)\phi_1(y_1)\cdots(\mathscr{b}_M(-\Delta)^{s_M}_{Y_M}\phi_M(y_M+e_M+\epsilon Y_M))\overline{\tau}_1(t_1)\cdots\overline{\tau}_l(t_l)=0,\end{aligned}$$

from which it follows that $w \in \mathcal{H}$.

In addition, using (5.13), (5.91), and (5.100), we have

$$\begin{aligned}0 = \theta\cdot\partial^K w(0) &= \sum_{|\iota|\le K}\theta_{i,I,\mathfrak{I}}\partial^i_x\partial^I_y\partial^{\mathfrak{I}}_t w(0) = \sum_{\substack{|\iota|\le K\\|I|\le|\overline{I}|}}\theta_{i,I,\mathfrak{I}}\partial^i_x\partial^I_y\partial^{\mathfrak{I}}_t w(0)\\ &= \sum_{\substack{|\iota|\le K\\|I|\le|\overline{I}|}}\theta_{i,I,\mathfrak{I}}\,\underline{x}^i\partial^{I_1}_{y_1}\phi_1(e_1+\epsilon Y_1)\cdots\partial^{I_M}_{y_M}\phi_M(e_M+\epsilon Y_M)\underline{t}^{\mathfrak{I}}.\end{aligned}$$

Hence, we set

$$\Xi := |\overline{I}| - \sum_{j=1}^{M} s_j,$$

we multiply the latter identity by ϵ^{Ξ}, and we exploit (5.98). In this way, we find

$$\begin{aligned}0 &= \lim_{\epsilon\searrow 0}\sum_{\substack{|\iota|\le K\\|I|\le|\overline{I}|}}\epsilon^{|\overline{I}|-|I|}\theta_{i,I,\mathfrak{I}}\,\underline{x}^i\epsilon^{|I_1|-s_1}\partial^{I_1}_{y_1}\phi_1(e_1+\epsilon Y_1)\cdots\epsilon^{|I_M|-s_M}\partial^{I_M}_{y_M}\phi_M(e_M+\epsilon Y_M)\underline{t}^{\mathfrak{I}}\\ &= \sum_{\substack{|\iota|\le K\\|I|=|\overline{I}|}}\theta_{i,I,\mathfrak{I}}\kappa_{s_j}\underline{x}^i e^I(-e_1\cdot Y_1)_+^{s_1-|I_1|}\cdots(-e_M\cdot Y_M)_+^{s_M-|I_M|}\underline{t}^{\mathfrak{I}}\\ &= (-e_1\cdot Y_1)_+^{s_1}\cdots(-e_M\cdot Y_M)_+^{s_M}\sum_{\substack{|\iota|\le K\\|I|=|\overline{I}|}}\theta_{i,I,\mathfrak{I}}\kappa_{s_j}\underline{x}^i e^I(-e_1\cdot Y_1)_+^{-|I_1|}\cdots(-e_M\cdot Y_M)_+^{-|I_M|}\underline{t}^{\mathfrak{I}},\end{aligned}$$

and consequently

$$0 = \sum_{\substack{|\iota|\le K\\|I|=|\overline{I}|}}\theta_{i,I,\mathfrak{I}}\kappa_{s_j}\underline{x}^i e^I(-e_1\cdot Y_1)_+^{-|I_1|}\cdots(-e_M\cdot Y_M)_+^{-|I_M|}\underline{t}^{\mathfrak{I}}. \tag{5.102}$$

Now we take free parameters $\underline{y} \in \mathbb{R}^{m_1+\dots+m_M}\setminus\{0\}$ and we perform the same change of variables as in (5.58). In this way, we deduce from (5.102) that

$$0 = \sum_{\substack{|i|+|I|+|\mathfrak{I}|\le K\\|I|=|\overline{I}|}} C_{i,I,\mathfrak{I}}\theta_{i,I,\mathfrak{I}}\,\underline{x}^i\underline{y}^I\underline{t}^{\mathfrak{I}},$$

for some $C_{i,I,\mathfrak{I}} \neq 0$, and the first claim in (5.15) is proved in this case as well. □

Proof of (5.15), *case 4.* Note that if there exists $j \in \{1, \dots, M\}$ such that $\mathfrak{b}_j \neq 0$, we are in the setting of case 3. Therefore, we assume that $\mathfrak{b}_j = 0$ for every $j \in \{1, \dots, M\}$.

We let ψ be the function constructed in Lemma 3.3. For each $h \in \{1, \dots, l\}$, we let $\overline{\psi}_h(t_h) := \psi(t_h)$, to make the dependence on h clear and explicit. Then, by formulas (3.8) and (3.9), we know that

$$D^{\alpha_h}_{t_h,0} \overline{\psi}_h(t_h) = 0 \quad \text{in } (1, +\infty) \tag{5.103}$$

and, for every $\ell \in \mathbb{N}$,

$$\lim_{\epsilon \searrow 0} \epsilon^{\ell - \alpha_h} \partial^{\ell}_{t_h} \overline{\psi}_h(1 + \epsilon t_h) = \kappa_{h,\ell} t_h^{\alpha_h - \ell}, \tag{5.104}$$

in the sense of distribution, for some $\kappa_{h,\ell} \neq 0$.

Now, we introduce a set of auxiliary parameters $\underline{t} = (\underline{t}_1, \dots, \underline{t}_l) \in (1, 2)^l$, and we fix ϵ sufficiently small and possibly depending on the parameters. Then, we define

$$a = (a_1, \dots, a_l) := \left(-\frac{\epsilon}{\underline{t}_1} - 1, \dots, -\frac{\epsilon}{\underline{t}_l} - 1 \right) \in (-2, 0)^l, \tag{5.105}$$

and

$$\psi_h(t_h) := \overline{\psi}_h(t_h - a_h). \tag{5.106}$$

With a simple computation we can show that the function in (5.106) satisfies

$$D^{\alpha_h}_{t_h,a_h} \psi_h(t_h) = D^{\alpha_h}_{t_h,0} \overline{\psi}_h(t_h - a_h) = 0 \quad \text{in } (1 + a_h, +\infty) = \left(-\frac{\epsilon}{\underline{t}_h}, +\infty \right), \tag{5.107}$$

thanks to (5.103). In addition, for every $\ell \in \mathbb{N}$, we have $\partial^{\ell}_{t_h} \psi_h(t_h) = \partial^{\ell}_{t_h} \overline{\psi}_h(t_h - a_h)$, and therefore, in light of (5.104) and (5.105),

$$\epsilon^{\ell - \alpha_h} \partial^{\ell}_{t_h} \psi_h(0) = \epsilon^{\ell - \alpha_h} \partial^{\ell}_{t_h} \overline{\psi}_h(-a_h) = \epsilon^{\ell - \alpha_h} \partial^{\ell}_{t_h} \overline{\psi}_h\left(1 + \frac{\epsilon}{\underline{t}_h}\right) \to \kappa_{h,\ell} \underline{t}_h^{\ell - \alpha_h}, \tag{5.108}$$

in the sense of distributions, as $\epsilon \searrow 0$.

Moreover, since for any $h = 1, \dots, l$, $\psi_h \in C^{k_h, \alpha_h}_{a_h}$, we can consider the extension

$$\psi_h^{\star}(t_h) := \begin{cases} \psi_h(t_h) & \text{if } t_h \in [a_h, +\infty), \\ \sum_{i=0}^{k_h - 1} \frac{\psi_h^{(i)}(a_h)}{i!} (t_h - a_h)^i & \text{if } t_h \in (-\infty, a_h), \end{cases} \tag{5.109}$$

and, using Lemma A.3 in [20] with $u := \psi_h$, $a := -\infty$, $b := a_h$, and $u_{\star} := \psi_h^{\star}$, we have

$$\psi_h^{\star} \in C^{k_h, \alpha_h}_{-\infty} \quad \text{and} \quad D^{\alpha_h}_{t_h,-\infty} \psi_h^{\star} = D^{\alpha_h}_{t_h,a_h} \psi_h = 0 \quad \text{in every interval } I \Subset \left(-\frac{\epsilon}{\underline{t}_h}, +\infty \right). \tag{5.110}$$

Now, we fix a set of free parameters $\underline{y} = (\underline{y}_1, \dots, \underline{y}_M) \in \mathbb{R}^{m_1+\dots+m_M}$, and we consider $\overline{\tau} \in C^\infty(\mathbb{R}^{m_1+\dots+m_M})$, such that

$$\overline{\tau}(y) := \begin{cases} \exp(\underline{y} \cdot y) & \text{if } y \in B_1^{m_1+\dots+m_M}, \\ 0 & \text{if } y \in \mathbb{R}^{m_1+\dots+m_M} \setminus B_2^{m_1+\dots+m_M}, \end{cases} \tag{5.111}$$

where

$$\underline{y} \cdot y = \sum_{j=1}^{M} \underline{y}_j \cdot y_j$$

denotes the standard scalar product.

We note that, for any multi-index $I \in \mathbb{N}^{m_1+\dots m_M}$,

$$\partial_y^I \overline{\tau}(0) = \underline{y}^I, \tag{5.112}$$

where the multi-index notation has been used.

Now, we define

$$w(x, y, t) := \tau(x)\overline{\tau}(y)\psi_1^\star(t_1)\cdots\psi_l^\star(t_l), \tag{5.113}$$

where the setting in (5.90), (5.109), and (5.111) has been exploited.

Using (5.110), we have, for any (x, y) in a neighborhood of the origin and $t \in (-\frac{\epsilon}{2}, +\infty)^l$,

$$\begin{aligned} \Lambda_{-\infty} w(x, y, t) &= \tau(x)\overline{\tau}(y)({}_{C_1}D^{\alpha_1}_{t_1,-\infty}\psi_1^\star(t_1))\cdots\psi_l^\star(t_l) \\ &\quad + \dots + \tau(x)\overline{\tau}(y)\psi_1^\star(t_1)\cdots({}_{C_l}D^{\alpha_l}_{t_l,-\infty}\psi_l^\star(t_l)) = 0. \end{aligned}$$

We have $w \in \mathcal{A}$, and, since τ and $\overline{\tau}$ are compactly supported, we also have $w \in C(\mathbb{R}^N) \cap C_0(\mathbb{R}^{N-l})$. Also, from Lemma 3.3, for any $h \in \{1, \dots, l\}$, we know that $\overline{\psi}_h \in C^\infty((1, +\infty))$, hence $\psi_h \in C^\infty((-\frac{\epsilon}{\underline{t}_h}, +\infty))$. Thus, $w \in C^\infty(\mathcal{N})$, and consequently $w \in \mathcal{H}$.

Recalling (5.14), (5.91), and (5.112), we have

$$\begin{aligned} 0 = \theta \cdot \partial^K w(0) &= \sum_{|\iota| \le K} \theta_{i,I,\mathfrak{J}} \partial_x^i \partial_y^I \partial_t^{\mathfrak{J}} w(0) = \sum_{\substack{|\iota| \le K \\ |\mathfrak{J}| \le |\overline{\mathfrak{J}}|}} \theta_{i,I,\mathfrak{J}} \partial_x^i \partial_y^I \partial_t^{\mathfrak{J}} w(0) \\ &= \sum_{\substack{|\iota| \le K \\ |\mathfrak{J}| \le |\overline{\mathfrak{J}}|}} \theta_{i,I,\mathfrak{J}}\, \underline{x}^i \underline{y}^I \partial_{t_1}^{\mathfrak{J}_1}\psi_1(0)\cdots\partial_{t_l}^{\mathfrak{J}_l}\psi_l(0). \end{aligned} \tag{5.114}$$

Hence, we set

$$\Xi := |\overline{\mathfrak{J}}| - \sum_{h=1}^{l} \alpha_h,$$

we multiply the identity in (5.114) by ϵ^{Ξ}, and we exploit (5.108). In this way, we find that

$$\begin{aligned} 0 &= \lim_{\epsilon \searrow 0} \sum_{\substack{|\iota|\le K \\ |\mathfrak{I}|\le|\overline{\mathfrak{I}}|}} \epsilon^{|\overline{\mathfrak{I}}|-|\mathfrak{I}|} \theta_{i,I,\mathfrak{I}} \underline{x}^{i} \underline{y}^{I} \epsilon^{\mathfrak{I}_1-\alpha_1} \partial_{t_1}^{\mathfrak{I}_1} \psi_1(0) \cdots \epsilon^{\mathfrak{I}_l-\alpha_l} \partial_{t_l}^{\mathfrak{I}_l} \psi_l(0) \\ &= \sum_{\substack{|\iota|\le K \\ |\mathfrak{I}|=|\overline{\mathfrak{I}}|}} \theta_{i,I,\mathfrak{I}} \kappa_{1,\mathfrak{I}_1} \cdots \kappa_{l,\mathfrak{I}_l} \underline{x}^{i} \underline{y}^{I} \underline{t}_1^{\mathfrak{I}_1-\alpha_1} \cdots \underline{t}_l^{\mathfrak{I}_l-\alpha_l} \\ &= \underline{t}_1^{-\alpha_1} \cdots \underline{t}_l^{-\alpha_l} \sum_{\substack{|\iota|\le K \\ |\mathfrak{I}|=|\overline{\mathfrak{I}}|}} \theta_{i,I,\mathfrak{I}} \kappa_{1,\mathfrak{I}_1} \cdots \kappa_{l,\mathfrak{I}_l} \underline{x}^{i} \underline{y}^{I} \underline{t}_1^{\mathfrak{I}_1} \cdots \underline{t}_l^{\mathfrak{I}_l}, \end{aligned}$$

and consequently

$$0 = \sum_{\substack{|\iota|\le K \\ |\mathfrak{I}|=|\overline{\mathfrak{I}}|}} \theta_{i,I,\mathfrak{I}} \kappa_{1,\mathfrak{I}_1} \cdots \kappa_{l,\mathfrak{I}_l} \underline{x}^{i} \underline{y}^{I} \underline{t}^{\mathfrak{I}},$$

and the second claim in (5.15) is proved in this case as well. □

5.3 Every function is locally $\Lambda_{-\infty}$-harmonic up to a small error, and completion of the proof of Theorem 5.1

In this section we complete the proof of Theorem 5.1 (which in turn implies Theorem 2.1 via Lemma 5.2). By standard approximation arguments we can reduce to the case in which f is a polynomial, and hence, by the linearity of the operator $\Lambda_{-\infty}$, to the case in which f is a monomial. The details of the proof are as follows.

5.3.1 Proof of Theorem 5.1 when f is a monomial

We prove Theorem 5.1 under the initial assumption that f is a monomial, that is,

$$f(x,y,t) = \frac{x_1^{i_1} \cdots x_n^{i_n} y_1^{I_1} \cdots y_M^{I_M} t_1^{\mathfrak{I}_1} \cdots t_l^{\mathfrak{I}_l}}{\iota!} = \frac{x^i y^I t^{\mathfrak{I}}}{\iota!} = \frac{(x,y,t)^{\iota}}{\iota!}, \tag{5.115}$$

where $\iota! := i_1! \ldots i_n! I_1! \ldots I_M! \mathfrak{I}_1! \ldots \mathfrak{I}_l!$ and $I_\beta! := I_{\beta,1}! \ldots I_{\beta,m_\beta}!$, $i_\chi! := i_{\chi,1}! \ldots i_{\chi,p_\chi}!$ for all $\beta = 1, \ldots M$ and $\chi = 1, \ldots, n$. To this end, we argue as follows. We consider $\eta \in (0,1)$, to be taken sufficiently small with respect to the parameter $\epsilon > 0$ which has been fixed in the statement of Theorem 5.1, and we define

$$\mathcal{T}_\eta(x,y,t) := (\eta^{\frac{1}{r_1}} x_1, \ldots, \eta^{\frac{1}{r_n}} x_n, \eta^{\frac{1}{2s_1}} y_1, \ldots, \eta^{\frac{1}{2s_M}} y_M, \eta^{\frac{1}{\alpha_1}} t_1, \ldots, \eta^{\frac{1}{\alpha_l}} t_l).$$

We also define

$$\gamma := \sum_{j=1}^{n} \frac{|i_j|}{r_j} + \sum_{j=1}^{M} \frac{|I_j|}{2s_j} + \sum_{j=1}^{l} \frac{\mathfrak{I}_j}{\alpha_j} \tag{5.116}$$

and

$$\delta := \min\left\{\frac{1}{r_1}, \dots, \frac{1}{r_n}, \frac{1}{2s_1}, \dots, \frac{1}{2s_M}, \frac{1}{\alpha_1}, \dots, \frac{1}{\alpha_l}\right\}. \tag{5.117}$$

We also take $K_0 \in \mathbb{N}$ such that

$$K_0 \geq \frac{\gamma + 1}{\delta} \tag{5.118}$$

and we let

$$K := K_0 + |i| + |I| + |\mathfrak{I}| + \ell = K_0 + |\iota| + \ell, \tag{5.119}$$

where ℓ is the fixed integer given in the statement of Theorem 5.1.

By Lemma 5.3, there exist a neighborhood $\mathcal{N}$ of the origin and a function $w \in C(\mathbb{R}^N) \cap C_0(\mathbb{R}^{N-l}) \cap C^\infty(\mathcal{N}) \cap \mathcal{A}$ such that

$$\Lambda_{-\infty} w = 0 \text{ in } \mathcal{N} \tag{5.120}$$

and such that

$$\begin{aligned}&\text{all the derivatives of } w \text{ in } 0 \text{ up to order } K \text{ vanish,}\\ &\text{with the exception of } \partial^\iota w(0) \text{ which equals } 1,\end{aligned} \tag{5.121}$$

with ι being as in (5.115). Recalling the definition of $\mathcal{A}$ on page 89, we also know that

$$\partial_{t_h}^{k_h} w = 0 \text{ in } (-\infty, a_h), \tag{5.122}$$

for suitable $a_h \in (-2, 0)$, for all $h \in \{1, \dots, l\}$.

In this way, setting

$$g := w - f, \tag{5.123}$$

we deduce from (5.121) that

$$\partial^\sigma g(0) = 0 \quad \text{for any } \sigma \in \mathbb{N}^N \text{ with } |\sigma| \leq K.$$

Accordingly, in $\mathcal{N}$ we can write

$$g(x, y, t) = \sum_{|\tau| \geq K+1} x^{\tau_1} y^{\tau_2} t^{\tau_3} h_\tau(x, y, t), \tag{5.124}$$

for some h_τ smooth in $\mathcal{N}$, where the multi-index notation $\tau = (\tau_1, \tau_2, \tau_3)$ has been used.

Now, we define

$$u(x, y, t) := \frac{1}{\eta^\gamma} w(\mathcal{T}_\eta(x, y, t)). \tag{5.125}$$

In light of (5.122), we note that $\partial^{k_h}_{t_h} u = 0$ in $(-\infty, a_h/\eta^{\frac{1}{\alpha_h}})$, for all $h \in \{1,\dots,l\}$, and therefore $u \in C(\mathbb{R}^N) \cap C_0(\mathbb{R}^{N-l}) \cap C^\infty(\mathcal{T}_\eta(\mathcal{N})) \cap \mathcal{A}$. We also claim that

$$\mathcal{T}_\eta([-1,1]^{N-l} \times (a_1,+\infty) \times \cdots \times (a_l,+\infty)) \subseteq \mathcal{N}. \tag{5.126}$$

To check this, let $(x,y,t) \in [-1,1]^{N-l} \times (a_1+\infty) \times \cdots \times (a_l,+\infty)$ and $(X,Y,T) := \mathcal{T}_\eta(x,y,t)$. Then we have $|X_1| = \eta^{\frac{1}{r_1}}|x_1| \le \eta^{\frac{1}{r_1}}$, $|Y_1| = \eta^{\frac{1}{2s_1}}|y_1| \le \eta^{\frac{1}{2s_1}}$, $T_1 = \eta^{\frac{1}{\alpha_1}} t_1 > a_1\eta^{\frac{1}{\alpha_1}} > -1$, provided η is small enough. Repeating this argument, we obtain, for small η,

$$(X,Y,T) \text{ is as close to the origin as we wish.} \tag{5.127}$$

From (5.127) and the fact that $\mathcal{N}$ is an open set, we infer that $(X,Y,T) \in \mathcal{N}$, and this proves (5.126).

Thanks to (5.120) and (5.126), we have, in $B_1^{N-l} \times (-1,+\infty)^l$,

$$\begin{aligned}
&\eta^{\gamma-1}\Lambda_{-\infty} u(x,y,t) \\
&= \sum_{j=1}^{n} \mathfrak{a}_j \partial^{r_j}_{x_j} w(\mathcal{T}_\eta(x,y,t)) + \sum_{j=1}^{M} \mathfrak{b}_j(-\Delta)^{s_j}_{y_j} w(\mathcal{T}_\eta(x,y,t)) + \sum_{j=1}^{l} \mathfrak{c}_j D^{\alpha_h}_{t_h,-\infty} w(\mathcal{T}_\eta(x,y,t)) \\
&= \Lambda_{-\infty} w(\mathcal{T}_\eta(x,y,t)) \\
&= 0.
\end{aligned}$$

These observations establish that u solves the equation in $B_1^{N-l} \times (-1+\infty)^l$ and u vanishes when $|(x,y)| \ge R$, for some $R > 1$, and thus the claims in (5.3) and (5.4) are proved.

Now we prove that u approximates f, as claimed in (5.5). For this, using the monomial structure of f in (5.115) and the definition of γ in (5.116), we have, in multi-index notation,

$$\frac{1}{\eta^\gamma} f(\mathcal{T}_\eta(x,y,t)) = \frac{1}{\eta^\gamma \iota!}(\eta^{\frac{1}{r}}x)^i(\eta^{\frac{1}{2s}}y)^I(\eta^{\frac{1}{\alpha}}t)^{\mathfrak{I}} = \frac{1}{\iota!}x^i y^I t^{\mathfrak{I}} = f(x,y,t). \tag{5.128}$$

Consequently, by (5.123), (5.124), (5.125), and (5.128),

$$\begin{aligned}
u(x,y,t) - f(x,y,t) &= \frac{1}{\eta^\gamma} g(\eta^{\frac{1}{r_1}}x_1,\dots,\eta^{\frac{1}{r_n}}x_n, \eta^{\frac{1}{2s_1}}y_1,\dots,\eta^{\frac{1}{2s_M}}y_M, \eta^{\frac{1}{\alpha_1}}t_1,\dots,\eta^{\frac{1}{\alpha_l}}t_l) \\
&= \sum_{|\tau|\ge K+1} \eta^{|\frac{\tau_1}{r}|+|\frac{\tau_2}{2s}|+|\frac{\tau_3}{\alpha}|-\gamma} x^{\tau_1} y^{\tau_2} t^{\tau_3} h_\tau(\eta^{\frac{1}{r}}x, \eta^{\frac{1}{2s}}y, \eta^{\frac{1}{\alpha}}t),
\end{aligned}$$

where a multi-index notation has been used, e. g., we have written

$$\frac{\tau_1}{r} := \left(\frac{\tau_{1,1}}{r_1},\dots,\frac{\tau_{1,n}}{r_n}\right) \in \mathbb{R}^n.$$

Therefore, for any multi-index $\beta = (\beta_1, \beta_2, \beta_3)$ with $|\beta| \le \ell$,

$$\begin{aligned}
&\partial^\beta(u(x,y,t) - f(x,y,t)) \\
&= \partial_x^{\beta_1}\partial_y^{\beta_2}\partial_t^{\beta_3}(u(x,y,t) - f(x,y,t)) \\
&= \sum_{\substack{|\beta_1'|+|\beta_1''|=|\beta_1| \\ |\beta_2'|+|\beta_2''|=|\beta_2| \\ |\beta_3'|+|\beta_3''|=|\beta_3| \\ |\tau|\ge K+1}} c_{\tau,\beta}\eta^{\kappa_{\tau,\beta}} x^{\tau_1-\beta_1'} y^{\tau_2-\beta_2'} t^{\tau_3-\beta_3'} \partial_x^{\beta_1''}\partial_y^{\beta_2''}\partial_t^{\beta_3''} h_\tau(\eta^{\frac{1}{r}}x, \eta^{\frac{1}{2s}}y, \eta^{\frac{1}{\alpha}}t),
\end{aligned} \tag{5.129}$$

where

$$\kappa_{\tau,\beta} := \left|\frac{\tau_1}{r}\right| + \left|\frac{\tau_2}{2s}\right| + \left|\frac{\tau_3}{\alpha}\right| - \gamma + \left|\frac{\beta_1''}{r}\right| + \left|\frac{\beta_2''}{2s}\right| + \left|\frac{\beta_3''}{\alpha}\right|,$$

for suitable coefficients $c_{\tau,\beta}$. Thus, to complete the proof of (5.5), we need to show that this quantity is small if so is η. To this aim, we use (5.117), (5.118), and (5.119) to see that

$$\begin{aligned}
\kappa_{\tau,\beta} &\ge \left|\frac{\tau_1}{r}\right| + \left|\frac{\tau_2}{2s}\right| + \left|\frac{\tau_3}{\alpha}\right| - \gamma \\
&\ge \delta(|\tau_1| + |\tau_2| + |\tau_3|) - \gamma \\
&\ge K\delta - \gamma \\
&\ge K_0\delta - \gamma \\
&\ge 1.
\end{aligned}$$

Consequently, we deduce from (5.129) that $\|u - f\|_{C^\ell(B_1^N)} \le C\eta$ for some $C > 0$. By choosing η sufficiently small with respect to ϵ, this implies the claim in (5.5). This completes the proof of Theorem 5.1 when f is a monomial.

5.3.2 Proof of Theorem 5.1 when f is a polynomial

Now, we consider the case in which f is a polynomial. In this case, we can write f as

$$f(x,y,t) = \sum_{j=1}^{J} c_j f_j(x,y,t),$$

where each f_j is a monomial, $J \in \mathbb{N}$, and $c_j \in \mathbb{R}$ for all $j = 1, \dots, J$.

Let

$$c := \max_{j\in\{1,\dots,J\}} c_j.$$

Then, by the work presented in Section 5.3.1, we know that the claim in Theorem 5.1 holds for each f_j, and so we can find $a_j \in (-\infty, 0)^l$, $u_j \in C^\infty(B_1^N) \cap C(\mathbb{R}^N) \cap \mathcal{A}$, and $R_j > 1$ such that $\Lambda_{-\infty} u_j = 0$ in $B_1^{N-l} \times (-1, +\infty)^l$, $\|u_j - f_j\|_{C^\ell(B_1^N)} \le \epsilon$, and $u_j = 0$ if $|(x,y)| \ge R_j$.

Hence, we set

$$u(x,y,t) := \sum_{j=1}^{J} c_j u_j(x,y,t),$$

and we see that

$$\|u - f\|_{C^\ell(B_1^N)} \le \sum_{j=1}^{J} |c_j| \|u_j - f_j\|_{C^\ell(B_1^N)} \le cJ\epsilon. \tag{5.130}$$

Also, $\Lambda_{-\infty} u = 0$ thanks to the linearity of $\Lambda_{-\infty}$ in $B_1^{N-l} \times (-1,+\infty)^l$. Finally, u is supported in B_R^{N-l} in the variables (x,y), i. e.,

$$R := \max_{j\in\{1,\dots,J\}} R_j.$$

This proves Theorem 5.1 when f is a polynomial (up to replacing ϵ with $cJ\epsilon$).

5.3.3 Proof of Theorem 5.1 for a general f

Now we deal with the case of a general f. To this end, we exploit Lemma 2 in [25] and we see that there exists a polynomial $\tilde{f}$ such that

$$\|f - \tilde{f}\|_{C^\ell(B_1^N)} \le \epsilon. \tag{5.131}$$

Then, applying the result already proven in Section 5.3.2 to the polynomial $\tilde{f}$, we can find $a \in (-\infty,0)^l$, $u \in C^\infty(B_1^N) \cap C(\mathbb{R}^N) \cap \mathcal{A}$, and $R > 1$ such that

$$\begin{aligned} &\Lambda_{-\infty} u = 0 \quad \text{in } B_1^{N-l} \times (-1,+\infty)^l, \\ &u = 0 \quad \text{if } |(x,y)| \ge R, \\ &\partial_{t_h}^{k_h} u = 0 \quad \text{if } t_h \in (-\infty, a_h), \text{ for all } h \in \{1,\dots,l\}, \\ \text{and} \quad &\|u - \tilde{f}\|_{C^\ell(B_1^N)} \le \epsilon. \end{aligned}$$

Then, recalling (5.131), we see that

$$\|u - f\|_{C^\ell(B_1^N)} \le \|u - \tilde{f}\|_{C^\ell(B_1^N)} + \|f - \tilde{f}\|_{C^\ell(B_1^N)} \le 2\epsilon.$$

Hence, the proof of Theorem 5.1 is complete. □

A Some applications

In this appendix we give some applications of the approximation results obtained and discussed in this book. These examples exploit particular cases of the operator Λ_a, namely, when $s \in (0,1)$ and Λ_a is the fractional Laplacian $(-\Delta)^s$ or the fractional heat operator $\partial_t + (-\Delta)^s$. Similar applications have been pointed out in [17, 4, 63].

Example A.1 (The classical Harnack inequality fails for s-harmonic functions). The Harnack inequality, in its classical formulation, says that if u is a nontrivial and nonnegative harmonic function in B_1, then, for any $0 < r < 1$, there exists $0 < c = c(n,r)$ such that

$$\sup_{B_r} u \le c \inf_{B_r} u. \tag{A.1}$$

The same result is not true for s-harmonic functions. To construct a counterexample, consider the smooth function $f(x) = |x|^2$, and, for a small $\epsilon > 0$, let $v = v_\epsilon$ be the function provided by Theorem 2.1, where we choose $\ell = 0$. Note that, if $x \in B_1 \setminus B_{r/2}$,

$$v(x) \ge f(x) - \|v - f\|_{L^\infty(B_1)} \ge \frac{r^2}{4} - \epsilon > \frac{r^2}{8},$$

provided ϵ is small enough, while

$$v(0) \le f(0) + \|v - f\|_{L^\infty(B_1)} \le \epsilon < \frac{r^2}{8}.$$

Hence, we have $v(0) < v(x)$ for any $x \in B_1 \setminus B_{r/2}$, and therefore the minimum of v in B_1 is attained in some point $\overline{x} \in \overline{B_{r/2}}$. Then, we define

$$u(x) := v(x) - v(\overline{x}).$$

Note that u is s-harmonic in B_1 since so is v. Also, $u \ge 0$ in B_1 by construction, and $u > 0$ in $B_1 \setminus B_{r/2}$. However, since $\overline{x} \in B_r$

$$\inf_{B_r} u = u(\overline{x}) = 0,$$

which implies that u cannot satisfies an inequality such as (A.1).

As a matter of fact, in the fractional case, the analogue of the Harnack inequality requires u to be nonnegative in the whole of $\mathbb{R}^n$, hence a "global" condition is needed to obtain a "local" oscillation bound; see, e. g., [40] and the references therein for a complete discussion of nonlocal Harnack inequalities.

Example A.2 (A logistic equation with nonlocal interactions). We consider the logistic equation taken into account in [17]

$$-(-\Delta)^s u + (\sigma - \mu u)u + \tau(J * u) = 0, \tag{A.2}$$

https://doi.org/10.1515/9783110664355-006

where $s \in (0,1]$, $\tau \in [0,+\infty)$, and σ, μ, J are nonnegative functions. The symbol $*$ denotes as usual the convolution product between J and u. Moreover, the convolution kernel J is assumed to be of unit mass and even, namely,

$$\int_{\mathbb{R}^n} J(x)\,dx = 1$$

and

$$J(-x) = J(x) \quad \text{for any } x \in \mathbb{R}^n.$$

In this framework, the solution u denotes the density of a population living in some environment $\Omega \subseteq \mathbb{R}^n$, while the functions σ and μ model the growing and dying effects on the population, respectively. The equation is driven by the fractional Laplacian that models a nonlocal dispersal strategy which has been observed experimentally in nature and may be related to optimal hunting strategies and adaptation to the environment stimulated by natural selection.

We state here a result which translates the fact that a population with a nonlocal strategy can plan the distribution of resources in a strategic region better than populations with a local one.

Namely, for fixed $\Omega = B_1$, one can find a solution of a slightly perturbed version of (A.2) in B_1, compactly supported in a larger ball B_{R_ϵ}, where $\epsilon \in (0,1)$ denotes the perturbation.

The strategic plan consists in properly adjusting the resources in $B_{R_\epsilon} \setminus B_1$ (that is, a bounded region in which the equation is not satisfied) in order to consume almost all the given resources in B_1.

The detailed statement goes as follows.

Theorem A.3. *Let $s \in (0,1)$ and $\ell \in \mathbb{N}$, $\ell \geq 2$. Assume that $\sigma, \mu \in C^\ell(\overline{B_2})$, with*

$$\inf_{\overline{B_2}} \mu > 0, \quad \inf_{\overline{B_2}} \sigma > 0.$$

For fixed $\epsilon \in (0,1)$, there exist a nonnegative function u_ϵ, $R_\epsilon > 2$, and $\sigma_\epsilon \in C^\ell(\overline{B_1})$ such that

$$\begin{aligned}
&(-\Delta)^s u_\epsilon = (\sigma_\epsilon - \mu u_\epsilon)u_\epsilon + \tau(J * u_\epsilon) \quad \text{in } B_1,\\
&u_\epsilon = 0 \quad \text{in } \mathbb{R}^n \setminus B_{R_\epsilon},\\
&\|\sigma_\epsilon - \sigma\|_{C^\ell(\overline{B_1})} \leq \epsilon,\\
&u_\epsilon \geq \mu^{-1}\sigma_\epsilon \quad \text{in } B_1.
\end{aligned}$$

Example A.4. Higher order nonlocal equations also appear naturally in several contexts; see, e. g., [18] for a nonlocal version of the Cahn–Hilliard phase coexistence model. Higher orders operators have also appeared in connection with logistic equations; see, e. g., [11]. In this spirit, we point out a version of Theorem A.3 which is new and relies on Theorem 2.1. Its content is that nonlocal logistic equations (of any order and with nonlocality given in time, space, or both) admits solutions which can arbitrarily well adapt to any given resource. The precise statement is the following.

Theorem A.5. *Let $s \in (0,+\infty)$, $\alpha \in (0,+\infty)$, and $\ell \in \mathbb{N}$, $\ell \geq 2$. Assume that*

$$\text{either } s \notin \mathbb{N} \text{ or } \alpha \notin \mathbb{N}. \tag{A.3}$$

Let $\sigma, \mu \in C^{\ell}(\overline{B_2})$, with

$$\inf_{\overline{B_1}} \mu > 0. \tag{A.4}$$

For fixed $\epsilon \in (0,1)$, there exist a nonnegative function u_ϵ, $R_\epsilon > 2$, $a_\epsilon < 0$, and $\sigma_\epsilon \in C^{\ell}(\overline{B_1})$ such that

$$\begin{aligned}
& D^{\alpha}_{t,a_\epsilon} u_\epsilon(x,t) + (-\Delta)^s u_\epsilon(x,t) = (\sigma_\epsilon(x,t) - \mu(x,t)u_\epsilon(x,t))u_\epsilon(x,t) \\
& \qquad \text{for all } (x,t) \in \mathbb{R}^p \times \mathbb{R} \text{ with } |(x,t)| < 1, && \text{(A.5)} \\
& u_\epsilon(x,t) = 0 \quad \text{if } |(x,t)| \geq R_\epsilon, && \text{(A.6)} \\
& \|\sigma_\epsilon - \sigma\|_{C^{\ell}(\overline{B_1})} \leq \epsilon, && \text{(A.7)} \\
& u_\epsilon = \mu^{-1}\sigma_\epsilon \geq \mu^{-1}\sigma - \epsilon \quad \text{in } B_1. && \text{(A.8)}
\end{aligned}$$

Proof. We use Theorem 2.1 in the case in which $\Lambda_a := D^{\alpha}_{t,a} + (-\Delta)^s$. Let $f := \sigma/\mu$. Then, by Theorem 2.1, which can be exploited here in view of (A.3), we obtain the existence of suitable u_ϵ, $R_\epsilon > 2$ and $a_\epsilon < 0$ satisfying (A.6),

$$\begin{aligned}
& D^{\alpha}_{t,a_\epsilon} u_\epsilon(x,t) + (-\Delta)^s u_\epsilon(x,t) = 0 \\
& \qquad \text{for all } (x,t) \in \mathbb{R}^p \times \mathbb{R} \text{ with } |(x,t)| < 1, && \text{(A.9)}
\end{aligned}$$

and

$$\|u_\epsilon - f\|_{C^{\ell}(\overline{B_1})} \leq \epsilon. \tag{A.10}$$

Then, we set $\sigma_\epsilon := \mu u_\epsilon$, and then, by (A.10),

$$\begin{aligned}
\|\sigma_\epsilon - \sigma\|_{C^{\ell}(\overline{B_1})} &\leq C\|\mu\|_{C^{\ell}(\overline{B_1})} \left\| \frac{\sigma_\epsilon}{\mu} - \frac{\sigma}{\mu} \right\|_{C^{\ell}(\overline{B_1})} \\
&= C\|\mu\|_{C^{\ell}(\overline{B_1})} \|u_\epsilon - f\|_{C^{\ell}(\overline{B_1})} \\
&\leq C\|\mu\|_{C^{\ell}(\overline{B_1})} \epsilon, && \text{(A.11)}
\end{aligned}$$

which gives (A.7), up to renaming ϵ.

Moreover, if $|(x,t)| < 1$,

$$(\sigma_\epsilon - \mu u_\epsilon)u_\epsilon = 0 = D^\alpha_{t,a_\epsilon} u_\epsilon + (-\Delta)^s u_\epsilon,$$

thanks to (A.9), and this proves (A.5).

In addition, recalling (A.11) and (A.4),

$$u_\epsilon = \mu^{-1}\sigma_\epsilon \geq \mu^{-1}\sigma - \frac{1}{\inf_{\overline{B_1}} \mu} \|\sigma - \sigma_\epsilon\|_{L^\infty(B_1)} \geq \mu^{-1}\sigma - \frac{\|\mu\|_{C^\ell(\overline{B_1})}\epsilon}{\inf_{\overline{B_1}} \mu},$$

in B_1, which proves (A.8), up to renaming ϵ. □

Bibliography

[1] Nicola Abatangelo, Sven Jarohs, Alberto Saldaña, *Positive powers of the Laplacian: From hypersingular integrals to boundary value problems*, Commun. Pure Appl. Anal., **17** (2018), no. 3, 899–922, DOI 10.3934/cpaa.2018045. MR3809107.

[2] Nicola Abatangelo, Sven Jarohs, Alberto Saldaña, *Green function and Martin kernel for higher-order fractional Laplacians in balls*, Nonlinear Anal., **175** (2018) 173–90, DOI 10.1016/j.na.2018.05.019. MR3830727.

[3] Nicola Abatangelo, Sven Jarohs, Alberto Saldaña, *Integral representation of solutions to higher-order fractional Dirichlet problems on balls*, Commun. Contemp. Math., **20** (2018), no. 8, 1850002, DOI 10.1142/S0219199718500025.

[4] Nicola Abatangelo, Enrico Valdinoci, *Getting acquainted with the fractional Laplacian*, Springer INdAM Series, 2019.

[5] Niels Henrik Abel, *Œuvres complètes. Tome I*, Éditions Jacques Gabay, Sceaux, 1992 (French). Edited and with a preface by L. Sylow and S. Lie; Reprint of the second (1881) edition. MR1191901.

[6] Mark Allen, Luis Caffarelli, Alexis Vasseur, *A parabolic problem with a fractional time derivative*, Arch. Ration. Mech. Anal., **221** (2016), no. 2, 603–30, DOI 10.1007/s00205-016-0969-z. MR3488533.

[7] Ricardo Almeida, Nuno R. O. Bastos, M. Teresa T. Monteiro, *Modeling some real phenomena by fractional differential equations*, Math. Methods Appl. Sci., **39** (2016), no. 16, 4846–55, DOI 10.1002/mma.3818. MR3557159.

[8] John Andersson, *Optimal regularity for the Signorini problem and its free boundary*, Invent. Math., **204** (2016), no. 1, 1–82, DOI 10.1007/s00222-015-0608-6. MR3480553.

[9] V. E. Arkhincheev, É. M. Baskin, *Anomalous diffusion and drift in a comb model of percolation clusters*, J. Exp. Theor. Phys., **73** (1991), 161–5.

[10] A. V. Balakrishnan, *Fractional powers of closed operators and the semigroups generated by them*, Pacific J. Math., **10** (1960), 419–37. MR0115096.

[11] Mousomi Bhakta, *Solutions to semilinear elliptic PDE's with biharmonic operator and singular potential*, Electron. J. Differential Equations **261** (2016), 17. MR3578282.

[12] Umberto Biccari, Warma, Mahamadi, Enrique Zuazua, *Local elliptic regularity for the Dirichlet fractional Laplacian*, Adv. Nonlinear Stud., **17** (2017), no. 2, 387–409, DOI 10.1515/ans-2017-0014. MR3641649.

[13] Svetlana I. Boyarchenko, Sergei I. Levendorskiĭ, *Non-Gaussian Merton-Black-Scholes theory*, Advanced Series on Statistical Science & Applied Probability, vol. 9, World Scientific Publishing Co., Inc., River Edge, NJ, 2002. MR1904936.

[14] Claudia Bucur, *Some observations on the Green function for the ball in the fractional Laplace framework*, Commun. Pure Appl. Anal., **15** (2016), no. 2, 657–99, DOI 10.3934/cpaa.2016.15.657. MR3461641.

[15] Claudia Bucur, *Local density of Caputo-stationary functions in the space of smooth functions*, ESAIM Control Optim. Calc. Var., **23** (2017), no. 4, 1361–80, DOI 10.1051/cocv/2016056. MR3716924.

[16] Claudia Bucur, Enrico Valdinoci, *Nonlocal diffusion and applications*, Lecture Notes of the Unione Matematica Italiana, vol. 20, Springer/Unione Matematica Italiana, Cham/Bologna, 2016. MR3469920.

[17] Luis Caffarelli, Serena Dipierro, Enrico Valdinoci, *A logistic equation with nonlocal interactions*, Kinet. Relat. Models, **10** (2017), no. 1, 141–70, DOI 10.3934/krm.2017006. MR3579567.

https://doi.org/10.1515/9783110664355-007

[18] Luis Caffarelli, Enrico Valdinoci, *A priori bounds for solutions of a nonlocal evolution PDE*, Analysis and numerics of partial differential equations, Springer INdAM Ser., vol. 4, Springer, Milan, 2013, pp 141–63, DOI 10.1007/978-88-470-2592-9_10. MR3051400.

[19] Michele Caputo, *Linear models of dissipation whose Q is almost frequency independent. II*, Fract. Calc. Appl. Anal., **11** (2008), no. 1, 4–14, Reprinted from Geophys. J. R. Astr. Soc. **13** (1967), no. 5, 529–539, MR2379269.

[20] Alessandro Carbotti, Serena Dipierro, Enrico Valdinoci, *Local density of Caputo-stationary functions of any order*, Complex Var. Elliptic Equ. (2019), DOI 10.1080/17476933.2018.1544631.

[21] Donatella Danielli, Nicola Garofalo, Arshak Petrosyan, Tung To, *Optimal regularity and the free boundary in the parabolic Signorini problem*, Mem. Amer. Math. Soc., **249** (2017), no. 1181, v + 103, DOI 10.1090/memo/1181. MR3709717.

[22] Eleonora Di Nezza, Giampiero Palatucci, Enrico Valdinoci, *Hitchhiker's guide to the fractional Sobolev spaces*, Bull. Sci. Math., **136** (2012), no. 5, 521–73, DOI 10.1016/j.bulsci.2011.12.004. MR2944369.

[23] Mario Di Paola, Francesco Paolo Pinnola, Massimiliano Zingales, *Fractional differential equations and related exact mechanical models*, Comput. Math. Appl., **66** (2013), no. 5, 608–20, DOI 10.1016/j.camwa.2013.03.012. MR3089369.

[24] Serena Dipierro, Hans-Christoph Grunau, *Boggio's formula for fractional polyharmonic Dirichlet problems*, Ann. Mat. Pura Appl. (4), **196** (2017), no. 4, 1327–44, DOI 10.1007/s10231-016-0618-z. MR3673669.

[25] Serena Dipierro, Ovidiu Savin, Enrico Valdinoci, *All functions are locally s-harmonic up to a small error*, J. Eur. Math. Soc. (JEMS), **19** (2017), no. 4, 957–66, DOI 10.4171/JEMS/684. MR3626547.

[26] Serena Dipierro, Ovidiu Savin, Enrico Valdinoci, *Local approximation of arbitrary functions by solutions of nonlocal equations*, J. Geom. Anal., **29** (2019), 1428–55, DOI 10.1007/s12220-018-0045-z.

[27] Serena Dipierro, Enrico Valdinoci, *A Simple Mathematical Model Inspired by the Purkinje Cells: From Delayed Travelling Waves to Fractional Diffusion*, Bull. Math. Biol., **80** (2018), no. 7, 1849–70, DOI 10.1007/s11538-018-0437-z. MR3814763.

[28] Hongjie Dong, Doyoon Kim, *On L_p-estimates for a class of non-local elliptic equations*, J. Funct. Anal., **262** (2012), no. 3, 1166–99, DOI 10.1016/j.jfa.2011.11.002. MR2863859.

[29] Fausto Ferrari, *Weyl and Marchaud Derivatives: A Forgotten History*, Mathematics, **6** (2018), no. 1, DOI 10.3390/math6010006.

[30] Gaetano Fichera, *Sul problema elastostatico di Signorini con ambigue condizioni al contorno*, Atti Accad. Naz. Lincei Rend. Cl. Sci. Fis. Mat. Natur. (8), **34** (1963) 138–42 (Italian). MR0176661.

[31] Nicola Garofalo, *Fractional thoughts*, arXiv e-prints (2017), available at 1712.03347.

[32] Filippo Gazzola, Hans-Christoph Grunau, Guido Sweers, *Polyharmonic boundary value problems*, Lecture Notes in Mathematics, vol. 1991, Springer-Verlag, Berlin, 2010. Positivity preserving and nonlinear higher order elliptic equations in bounded domains. MR2667016.

[33] Tuhin Ghosh, Mikko Salo, Gunther Uhlmann, *The Calderón problem for the fractional Schrödinger equation*, ArXiv e-prints (2016), available at 1609.09248.

[34] David Gilbarg, Neil S. Trudinger, *Elliptic partial differential equations of second order*, Classics in Mathematics, Springer-Verlag, Berlin, 2001. Reprint of the 1998 edition. MR1814364.

[35] Tepper L. Gill, Woodford W. Zachary, *Functional analysis and the Feynman operator calculus*, Springer, Cham, 2016. MR3468941.

[36] Rudolf Gorenflo, Anatoly A. Kilbas, Francesco Mainardi, Sergei V. Rogosin, *Mittag-Leffler functions, related topics and applications*, Springer Monographs in Mathematics, Springer,

Heidelberg, 2014. MR3244285.

[37] Markus Haase, *The functional calculus for sectorial operators*, Operator Theory: Advances and Applications, vol. 169, Birkhäuser Verlag, Basel, 2006. MR2244037.

[38] Miguel de Icaza Herrera, *Galileo, Bernoulli, Leibniz and Newton around the brachistochrone problem*, Rev. Mexicana Fís., **40** (1994), no. 3, 459–75 (English, with English and Spanish summaries). MR1281370.

[39] S. L. Kalla, B. Ross, *The development of functional relations by means of fractional operators*, Fractional calculus (Glasgow, 1984) Res. Notes in Math., vol. 138, Pitman, Boston, MA, 1985, pp. 32–43, MR860085.

[40] Moritz Kassmann, *A new formulation of Harnack's inequality for nonlocal operators*, C. R. Math. Acad. Sci. Paris, **349** (2011), no. 11–12, 637–40, DOI 10.1016/j.crma.2011.04.014 (English, with English and French summaries). MR2817382.

[41] Anatoly A. Kilbas, Hari M. Srivastava, Juan J. Trujillo, *Theory and applications of fractional differential equations*, North-Holland Mathematics Studies, vol. 204, Elsevier Science B.V., Amsterdam, 2006. MR2218073.

[42] Nicolai V. Krylov, *On the paper "All functions are locally s-harmonic up to a small error" by Dipierro, Savin, and Valdinoci*, ArXiv e-prints (2018), available at 1810.07648.

[43] Rafael de la Llave, Enrico Valdinoci, *L^p-bounds for quasi-geostrophic equations via functional analysis*, J. Math. Phys., **52** (2011), no. 8, 083101, DOI 10.1063/1.3621828. MR2858052.

[44] Jesper Lützen, *Heaviside's operational calculus and the attempts to rigorise it*, Arch. Hist. Exact Sci., **21** (1979/80), no. 2, 161–200, DOI 10.1007/BF00330405. MR555103.

[45] Francesco Mainardi, *Fractional calculus and waves in linear viscoelasticity*, An introduction to mathematical models, Imperial College Press, London, 2010. MR2676137.

[46] Annalisa Massaccesi, Enrico Valdinoci, *Is a nonlocal diffusion strategy convenient for biological populations in competition?*, J. Math. Biol., **74** (2017), no. 1–2, 113–47, DOI 10.1007/s00285-016-1019-z. MR3590678.

[47] Benoit B. Mandelbrot, John W. Van Ness, *Fractional Brownian motions, fractional noises and applications*, SIAM Rev., **10** (1968) 422–37, DOI 10.1137/1010093. MR0242239.

[48] Benoit Mandelbrot, *The variation of certain speculative prices [reprint of J. Bus.* ***36*** *(1963), no. 4, 394–419]*, Financial risk measurement and management, Internat. Lib. Crit. Writ. Econ., vol. 267, Edward Elgar, Cheltenham, 2012, pp. 230–55. MR3235230.

[49] Kenneth S. Miller, Bertram Ross, *An introduction to the fractional calculus and fractional differential equations*, A Wiley-Interscience Publication, John Wiley & Sons, Inc., New York, 1993. MR1219954.

[50] Jan Mikusiński, *Operational calculus*, International Series of Monographs on Pure and Applied Mathematics, vol. 8, Pergamon Press/Państwowe Wydawnictwo Naukowe, New York–London–Paris–Los Angeles/Warsaw, 1959. MR0105594.

[51] Junichi Nakagawa, Kenichi Sakamoto, Masahiro Yamamoto, *Overview to mathematical analysis for fractional diffusion equations—new mathematical aspects motivated by industrial collaboration*, J. Math-for-Ind., **2A** (2010) 99–108, MR2639369.

[52] Keith B. Oldham, Jerome Spanier, *The fractional calculus*, Academic Press [A subsidiary of Harcourt Brace Jovanovich, Publishers], New York–London, 1974. Theory and applications of differentiation and integration to arbitrary order; With an annotated chronological bibliography by Bertram Ross; Mathematics in Science and Engineering, vol. 111. MR0361633.

[53] E. Omey, E. Willekens, *Abelian and Tauberian theorems for the Laplace transform of functions in several variables*, J. Multivariate Anal., **30** (1989), no. 2, 292–306, DOI 10.1016/0047-259X(89)90041-9. MR1015374.

[54] J. N. Pandey, *The Hilbert transform of Schwartz distributions and applications*, Pure and Applied Mathematics (New York), John Wiley & Sons, Inc., New York, 1996. A Wiley-Interscience

Publication. MR1363489.

[55] Arshak Petrosyan, Henrik Shahgholian, Nina Uraltseva, *Regularity of free boundaries in obstacle-type problems*, Graduate Studies in Mathematics, vol. 136, American Mathematical Society, Providence, RI, 2012. MR2962060.

[56] Igor Podlubny, *Fractional differential equations*, Mathematics in Science and Engineering, vol. 198, Academic Press, Inc., San Diego, CA, 1999. An introduction to fractional derivatives, fractional differential equations, to methods of their solution and some of their applications. MR1658022.

[57] Benjamin M. Regner, Dejan Vučinić, Cristina Domnisoru, Thomas M. Bartol, Martin W. Hetzer, Daniel M. Tartakovsky, Terrence J. Sejnowski, *Anomalous diffusion of single particles in cytoplasm*, Biophys. J. **104** (2013), no. 8, 1652–60, DOI 10.1016/j.bpj.2013.01.049.

[58] Bertram Ross, *The development, theory and applications of the Gamma-function and a profile of fractional calculus*, ProQuest LLC, Ann Arbor, MI, 1974. Thesis (Ph.D.)–New York University. MR2624107.

[59] Bertram Ross, *The development of fractional calculus 1695–1900*, Historia Math., **4** (1977) 75–89, DOI 10.1016/0315-0860(77)90039-8 (English, with German and French summaries). MR0444394.

[60] Bertram Ross, *Origins of fractional calculus and some applications*, Internat. J. Math. Statist. Sci., **1** (1992), no. 1, 21–34, MR1251624.

[61] Angkana Rüland, *Quantitative invertibility and approximation for the truncated Hilbert and Riesz Transforms*, ArXiv e-prints (2017), available at 1708.04285.

[62] Angkana Rüland, Mikko Salo, *The fractional Calderón problem: low regularity and stability*, ArXiv e-prints (2017), available at 1708.06294.

[63] Angkana Rüland, Mikko Salo, *Quantitative approximation properties for the fractional heat equation*, ArXiv e-prints (2017), available at 1708.06300.

[64] Angkana Rüland, Mikko Salo, *Exponential instability in the fractional Calderón problem*, Inverse Problems, **34** (2018), no. 4, 045003, DOI 10.1088/1361-6420/aaac5a. MR3774704.

[65] Stefan G. Samko, Anatoly A. Kilbas, Oleg I. Marichev, *Fractional integrals and derivatives*, Gordon and Breach Science Publishers, Yverdon, 1993. Theory and applications; Edited and with a foreword by S. M. Nikol'skiĭ; Translated from the 1987 Russian original; Revised by the authors. MR1347689.

[66] F. Santamaria, S. Wils, E. De Schutter, G. J. Augustine, *Anomalous diffusion in Purkinje cell dendrites caused by spines*, Neuron. **52** (2006), no. 4, 635–48. DOI 10.1016/j.neuron.2006.10.025.

[67] H. Schiessel, A. Blumen, *Hierarchical analogues to fractional relaxation equations*, J. Phys. A: Math. Gen. **26** (1993), no. 19, 5057–69. DOI 10.1088/0305-4470/26/19/034.

[68] A. Signorini, *Questioni di elasticità non linearizzata e semilinearizzata*, Rend. Mat. e Appl. (5), **18** (1959) 95–139 (Italian), MR0118021.

[69] Chung-Sik Sin, Liancun Zheng, *Existence and uniqueness of global solutions of Caputo-type fractional differential equations*, Fract. Calc. Appl. Anal., **19** (2016), no. 3, 765–74, DOI 10.1515/fca-2016-0040. MR3563609.

[70] William Seffens, *Models of RNA interaction from experimental datasets: framework of resilience*, Applications of RNA-Seq and Omics Strategies (2017), DOI 10.5772/intechopen.69452.

[71] P. R. Stinga, *User's guide to the fractional Laplacian and the method of semigroups*, arXiv e-prints (2018), available at 1808.05159.

[72] E. C. Titchmarsh, *Introduction to the theory of Fourier integrals*, 33rd ed., Chelsea Publishing Co., New York, 1986. MR942661.

[73] Enrico Valdinoci, *From the long jump random walk to the fractional Laplacian*, Bol. Soc. Esp.

Mat. Apl. SeMA, **49** (2009), 33–44, MR2584076.

[74] G. M. Viswanathan, V. Afanasyev, S. V. Buldyrev, E. J. Murphy, P. A. Prince, H. E. Stanley, *Lévy flight search patterns of wandering albatrosses*, Nature, **381** (1996) 413–5 DOI 10.1038/381413a0.

Index

De Gruyter Studies in Mathematics

Volume 41
Ulrich Knauer, Kolja Knauer
Algebraic Graph Theory. Morphisms, Monoids and Matrices, 2nd Edition, 2019
ISBN 978-3-11-061612-5, e-ISBN 978-3-11-061736-8,
e-ISBN (ePUB) 978-3-11-061628-6

Volume 43
Mark M. Meerschaert, Alla Sikorskii
Stochastic Models for Fractional Calculus, 2nd Edition, 2019
ISBN 978-3-11-055907-1, e-ISBN 978-3-11-056024-4,
e-ISBN (ePUB) 978-3-11-055914-9

Volume 52
Miroslav Pavlović
Function Classes on the Unit Disc, 2nd Edition, 2019
ISBN 978-3-11-062844-9, e-ISBN 978-3-11-063085-5,
e-ISBN (ePUB) 978-3-11-062865-4

Volume 73
Carol Jacoby, Peter Loth
Abelian Groups: Structures and Classifications, 2019
ISBN 978-3-11-043211-4, e-ISBN 978-3-11-042768-4,
e-ISBN (ePUB) 978-3-11-042786-8

Volume 72
Francesco Aldo Costabile
Modern Umbral Calculus: An Elementary Introduction with Applications to Linear Interpolation and Operator Approximation Theory, 2019
ISBN 978-3-11-064996-3, e-ISBN 978-3-11-065292-5,
e-ISBN (ePUB) 978-3-11-065009-9

Volume 34/1
József Lörinczi, Fumio Hiroshima, Volker Betz
Feynman-Kac-Type Theorems and Gibbs Measures: Volume 1, 2019
ISBN 978-3-11-033004-5, e-ISBN 978-3-11-033039-7,
e-ISBN (ePUB) 978-3-11-038993-7

Bei Fragen zur Produktsicherheit wenden Sie sich bitte an:
If you have any questions regarding product safety,
please contact:

Walter de Gruyter GmbH
Genthiner Straße 13
10785 Berlin
productsafety@degruyterbrill.com